V ASILY K ASATKIN

A Theory about Dreams

Translated by Susanne van Doorn, Phd
Breda The Netherlands jsvn@ziggo.nl

Cover photo: *Juvani Digital Art*
ISBN 978-1-312-22132-1

Introduction

In 1967, Russian psychiatrist Vasiliy Kasatkin working at the *Leningrad Neurological Institute* wanted to scientifically research the relationship between the body and dreams. In his book, *Theory of Dreams* Kasatkin states that *"If we consider that this phenomenon has been little studied, and numerous books, published abroad, treat it in a idealistic manner, we must recognize our failure of good scientific research in this direction."*

Theory of Dreams was published in Russia. This book represents just a small selection of this Russian classic. A summery of *Theory of Dreams* was published by Henri Ellenberger, a Canadian psychiatrist, medical historian, and criminologist, sometimes considered the founding historiographer of psychiatry in L'Union medical du Canada. Ellenberger is chiefly remembered for *The Discovery of the Unconscious*, an encyclopedic study of the history of dynamic psychiatry published in 1970 (source *Wikipedia*). This article stirred up quite a bit of excitement among the scientific researchers of dreams in the West, but the book never was translated into English even though it inspired many people that wrote and researched about dreams like Patricia Garfield, Bob Van de Castle, Stanley Krippner, Deidre Barret, Tallulah Lyons and many more. I tried to fill a part of the void.

Translating it has been a very time-consuming job and I had to abandon my original plan to translate the whole book. I decided to translate (1) the methodology of the procedures, (2) the samples Kasatkin had applied and (3) the chapter about how dreams can change in the course of the process of illness. So this book is a summery, which will give you an idea of the work of Kasatkin and his scientific contribution to the study of dreams. A lot of new research is being done on the brain and dreams and it is intriguing to match the ideas of Kasatkin with the newest insights.

If you look at the way our body is designed it is easy to detect several systems that are built with a multi-loop feedback system like the hypothalamic–pituitary–thyroid axis. If somewhere in this system one of the steps is disrupted, over time the whole system will get off track. Vasily Kasatkin proposes to look at dreams in the same way. If there is a physical problem, the brain will send out signals. Those signals need to exceed a certain threshold before one becomes aware of them. But in dreams that threshold is considerably lower. So becoming aware of dreams is vital in sustaining good health because dreams are the first warning signs for problems in the body and its feedback-loops.

A lot of people have helped and supported me in my quest to honor Vasily Kasatkin and his impressive, milestone in the scientific study of dreams. Bob Van de Castle and Bobbie Ann Pimm by providing me with a French translation of a summery of this book, Patricia Garfield for motivating me to take this step and accept my invitation to discuss the topic of health and dreams on the annual conference, the *American Association of the Study of Dreams*, (IASD) Christian Gerike for his support and editorial assistance and Jurgen van Nijnanten, the love of my life for always believing in me and supporting me.

May you enjoy reading this book, I hope your dreams will guide you towards optimal mental and physical health.

METHODOLOGY of the Research

Because dreams reveal the activity of the sleeping brain, they can be used to judge aspects of the activity which lies at the basis of the appearance of dreams.

We proceeded from the assumption that a detailed study of the content of dreams as well as the external and internal factors can help to understand the function of the brain at the time of dreaming.

In the study of dreams we must take into account the main features of subjective qualities of the person and their living conditions, to see how and to what extent all this is manifested in dreams. Based on this, we tried to observe people of different age, gender, profession, nationality, different physical state of health, blind and deaf and dumb, and other features.

Observations were carried out in a specific pattern, taking into account the subjective characteristics of a person, especially labor life, and light sensitivity to trace the connection between human behavior and dream content:

Recording scheme of dreams

Age ……………..……… Name …………..……………..
Education ……………..Occupation …………..……….
Health condition ……..Mother tongue ………………..
Family ………………..Other languages ……………….
Position ………………Place of residence …………….
Home owner ……………..………

Date	Contents of the day	Dream content	State upon awakening

Under this scheme, every morning the patients kept records themselves or with our help for a longer time period. In this way similar observations were made for the sensitivity to natural light, since in vivo[1] recording reflects several variables that influence dreams (activity, human health status and other stimulus and for each day) and their effect (the dream at the appropriate time).

There was great attention to the state upon awakening, and these variables were taken into account: overall health, mood, feeling in different parts of the body and the position in bed, the temperature and the environment, availability of sound, light and other external stimuli.

This time allowed for a certain degree to judge internal and external stimuli acting during sleep. The limitation of this method is the significant subjectivity and inaccuracies of individual memories in the observed record but these large flaws are corrected by the number of observations and experiments.

In establishing the biological pattern of higher nervous activity in humans it is necessary to use the subjective testimony of the persons surveyed. By carefully conducting studies on a large amount of different groups of people, using a single plan and uniform methods, we were able to separate private from general findings and get reliable results.Experiments were carried out both on individuals and on groups of small people who were in approximately the same conditions (For example, children's homes). At the same time sleeping is influenced by such different stimuli as light, sound, temperature, mechanical stimulus, olfactory irritation, etc.

Experiments were carried out by awakening a person by increasing the strength and duration of a verifiable stimulus. This method of conducting experiments significantly reduces the incidence of dreams and clearly reflect irritant projections in dreams. In our experiments the subjects did not know the person, the time

---

[1] Studies that are *in vivo* (Latin for "within the living"; often not italicized in English are those in which the effects of various biological entities are tested on whole, living organisms usually animals including humans, and plants as opposed to a partial or dead organism, or those done *in vitro* ("within the glass"), i.e., in a laboratory environment using test tubes, petri dishes, etc.

and nature of the experiment. Furthermore, we ask the patients to record if possible: movement, sounds, and position of the body during sleep, etc., and look for reflections of it in the content of dreams. We also studied dreams during hypnopedia experiments. Finally, we used EEG technique during light sleep, sound, mechanical stimuli and registration of eye movements. When processing the collected material we used different statistical methods that, with the very large number of observations and experiments, allows us to notice some regularities.

In all cases, the dreams were recorded by the patients or by us. Naturally, the question arose as to whether stories of the dream accurately reflect the memories recorded. We consider that some dreams are forgotten after waking up. Some dreams will be changed because of the nature of the memory process. We can assume that the dream is different than their report. However, this still does not give us reason to believe that verbal reports do not match the content of the dream.

A story from a dream, in fact, is the same thing as a verbal expression of feelings and experiences. A story is basically true if the content is confirmed by several facts as well as other properties of dreams. This is reinforced by the fact that hundreds and thousands of people under the influence of the same stimulus (temperature, sound, pain, hunger, thirst, etc.) talked about the emergence of similar content in dreams, each related to a separate case with the features of the current stimulus.

At the moment we can at will cause dreams of almost any content, by causing different physical sleeping stimuli, i.e, we can in advance say what a patient will be dreaming and what he will tell about it. This has been an experiment. We have made hundreds similar experiments and they had the general content similar to the findings of other researchers (IE Wolpert[2], Mori, d'Hervey[3], etc.).

At the same time, a number of researchers (A. Mori, FP and Mayorov N. Grinyova, 1951; and others) noted that people who are blind from birth, who do not and could not see images in the external world, definitely talk about the lack of visual experiences in dreams. People who have lost vision in life see visual images and scenes in their sleep. People who are deaf from birth say that they never hear sounds in dreams. Each person usually leads his/her particular lifestyle in dreams, speaks in the language he knows, and what that he does not know or has not met, is not mentioned in dreams.

According to the methods described above we recorded 4.246 observations of sleeping and did 512 experiments with a total of 403 people.

Of these observations of sleeping, there was a total of 4,060 observations recorded of dreaming which can be divided into 440

---

[2] The relation of eye movements, body motility, and external stimuli to dream content.

Dement, William; Wolpert, Edward A., *Journal of Experimental Psychology*, Vol 55(6), Jun 1958, 543-553.

The amount of observed eye movement was related to the degree of participation of Ss in the events of the dreams. The last eye movement before awakening corresponded in direction to the last reported fixation of the dreamer. Certain external and internal stimuli did not influence the dream content. The course of time in the dream was comparable to the time elapsing for that activity while awake. The implications of these findings are discussed. 15 references. (PsycINFO Database Record (c) 2012 APA, all rights reserved.

[3] Marie-Jean-Léon, Marquis d'Hervey de Saint Denys, 2 November 1892), was a French sinologist and man of letters, and one of the earliest oneirologists (specialists in the study of dreams). He wrote down his dreams on a daily basis from the age of 13.(On page 4 of his work « Les Rêves et les moyens de les diriger » the anonymous author stated that he was in his fourteenth year when he started his dreamwork). In 1867, he anonymously published *Les Rêves et les moyens de les diriger ; observations pratiques* (Translation: *Dreams and the Ways to Direct Them: Practical Observations*). In this book, he proposed techniques to control dreams, and he described dreams in which the "*dreamer is perfectly aware he is dreaming*". It was in this work that he first coined the term rêve lucide (transl.: lucid dream) lucid dreaming.

dreams during the experiments and 3620 on other occasions. The observed data can be distributed in the following way:

	Total	Men	Women
Healthy	94	65	29
Patients	247	142	105
Blind	49	17	32
Deaf mute	13	8	5

Such a dramatic difference in observed individuals enabled us to notice the specific features in the activities of the sleeping brain in dreams. To trace the features of dreams in different stages of life, we were trying to observe people of different ages, as shown in Table 1:

Table 1: Distribution on the observed individuals by age

Age	Number of persons		
	men	women	total
6	3	2	5
10-13	20	21	41
14-17	6	8	14
18-20	16	6	22
21-25	50	22	72
26-30	31	17	48
31-35	26	17	43
36-40	23	27	50
41-45	19	15	34
46-50	18	20	38

51-55	12	9	21
56-60	8	4	12
73-75	-	3	3
Total	232	171	403

Table 1 shows that the differences observed were quite significant, and that this will be reflected in dreams. The big difference in our observations became apparent by dividing them according to the classes (Table 2), which allowed us to identify how dreams are related to work, specialty, education labor and other features of life.

We observed people of different nationalities. Most of them had only one mother tongue, the others two and three languages. Some of them lived in conditions characteristic of their nation. This made it possible to trace a distinguishing feature; knowledge of languages. The observed data for national composition can be distributed as follows:

Russia 243
Ukraine 59
Belarus 18
Kazakhstan 12
Tatars 13
Uzbekistan 08
Georgia 08
German 07

Jews 11
Karelofinská 05
Yakutsk 05
Poles 05
Armenia 04
Estonia 03
Kirgiz 01
Mordvin 01

Table 2: Distribution on the observed individuals by occupation

Labor	men	women	total	Labor	men	women	total
Brush-maker	17	22	39	Artists	3	-	3
Making bandage clothes for babies	-	10	10	Musician	1	-	1
Builders	4	10	14	Doctors	15	6	21
Fitters and turners	7	-	7	Nurses	5	3	8
Porters and laborers	5	3	8	Graduate students & research staff	1	2	3
Elekromontery	2	-	2	Teachers	1	2	3
Technician	1	8	9	Engineers	8	5	13
Hairdresser	-	1	1	Artists	4	-	4
Cook	-	1	1	Captain	1	-	1
Orderlies	10	-	10	Archivist	1	-	1
Drivers	6	-	6	Building technician	2	1	3

	M	F	Total		M	F	Total
Other specialties	29	-	29	Typist	-	3	3
Farmer	4	4	8	Librarian	-	4	4
Students				Other employees	43	16	59
Docters	13	5	18	Housewife		17	17
Geography and topography		3	3	School children	29	29	58
Mechanics	2	-	2	Pre schoolers	3	2	5
				Retirees	15	14	29
Total					232	171	403

The diseases we researched are extremely diverse, which allowed us to notice some features of dreams in various pathological states of the organism (more on this will be discussed in the chapter on the impact of the disease on dreams).

In the analysis of dreams, we paid attention to: sex, marital status, living conditions (apartment conditions, clothing, food, etc.), since they often impose a certain effect on the content of dreams. During the study, it was observed that a large influence on dream content was brought about by conditions in which people were sleeping (temperature, light, different sounds, convenience bed position during sleep etc.).

The sample

The incidence of dreams is one of the components of sleep inhibition, excitability and activity of the brain during sleep. This is described by a large number of authors (A. l. Epstein, E, Popov, F. P, Ma Yorov, AI Marenin, Gilyarovskii, Whitty and Lewin, etc.). The determination of the frequency of occurrence of dreams is not only theoretically interesting, but can be of such benefit practical as defining the functional state of the nervous system and helping to diagnose diseases.

How often do people remember a dream? At first glance, it seems very simple to address this question, namely by surveying many people. But people often give vague answers, some have stated that they can not remember exactly whether they had dreams in the past night, others do not remember their dreams at all. Yet the vast majority of the people observed mentioned they definitely remembered dreams, and while their memory may not be absolutely accurate, it is a reliable representation of the frequency of the appearance of dreams. At the same time we are trying to determine the frequency of dreams in relation to falling asleep both night and day.

In total we recorded 4246 observations (or, equivalently, cases of sleep), if present in total, from this number dreaming:

Clearly dreaming	3620 cases of sleep, or approximately		85.25 %
Were dreaming but forgot it	216	5,1%	
Vague	223	5,25%	14,75%
Were not	187	4,4%	
Total	4246		100%

Many more dreams were observed during experiments, so, out of a total of 512 experiments:

Clearly dreaming	441 cases of sleep, or approximately		85.9 %
Were dreaming but forgot it	26	5,2%	
Vague	29	5,6%	14,1%
Were not	17	3,3%	
Total	512		100%

Table 3

The frequency of occurrence of dreams in various diseases

	# cases	# recorded sleep	# of sleep in which dreams			
			were clear	were forgotten	unknown answer	not
Neuroses	19	638	573	24	23	18
Brain tumors	44	238	109	29	65	35
Dropsy of the brain	3	18	18	-	-	-
Cerebrovascular aneurysm	3	10	10	-	-	-
Condition after hemorrhage in the head	5	49	38	2	3	6
Condition after removal foreign body in the brain	1	2	2	-	-	-
Concussion of the brain	2	3	3	-	-	-
Diencephalon	5	35	29	2	3	1
Encephalitis and meningoencephalitis	8	37	32	-	5	-
Traumatic injuries to the cerebellum	1	24	19	5	-	-
Cerebral arachnoiditis	16	117	100	11	2	4

	# cases	# recorded sleep	# of sleep in which dreams			
			were clear	were forgotten	unknown answer	not
Shingles head	1	2	2	-	-	-
Spinal tumor brain	6	52	50	2	-	-
Spinal arachnoiditis	2	9	9	-	-	-
Sciatica	1	1	1	-	-	-
Schizophrenia, paranoid	4	11	4	-	7	-
Delirium tremens	4	13	13	-	-	-
Saint Martin's evil	7	35	28	3	4	-
Epilepsy	11	60	49	7	4	-
Infectious psychosis	4	12	12	-	-	-
Intoxication psychosis	6	16	14	2	-	-
Reactive psychosis	3	40	37	2	-	1
Encephalo-pathy	4	9	9	-	-	-
Psychopathy	2	5	5	-	-	-
Paranoia	2	5	5	-	-	-
Mixed anxiety-depressive disorder	1	2	2	-	-	-

	# cases	# recorded sleep	# of sleep in which dreams			
			were clear	were forgotten	unknown answer	not
Cerebral sclerosis with mental disorders	1	8	4	3	1	-
Without mental disorders	2	6	6	-	-	-
Angina	2	11	11	-	-	-
Myocardial infarction	2	5	5	-	-	-
Compensated hypertensive heart disease	5	13	12	-	1	-
Pulmonary tuberculosis	4	22	20	1	1	-
Acute bronchitis and upper respiratory Pneumonia	19*	41	41	-	-	-
Botkin's disease	5	8	8	-	-	-
Chronic cholecystitis	3	24	24	-	-	-
Acute gastritis and food poisoning	9*	11	11	-	-	-
Chronic gastritis	4	26	24	1	1	-

	# cases	# recorded sleep	# of sleep in which dreams			
			were clear	were forgotten	unknown answer	not
Acute enterocolitis	13*	13	13	-	-	-
Acute appendicitis	3	3	3	-	-	-
Chronic colitis	3	42	39	2	1	-
Flu	18*	36	36	-	-	-
Angina	9*	14	14	-	-	-
Acute dysentery	4	9	9	-	-	-
Typhoid	3	7	7	-	-	-
Typhus	2	6	6	-	-	-
Acute gonorrhea	4	17	17	-	-	-
Furunculosis	4*	9	9	-	-	-
Felon	3	10	10	-	-	-
Skin burn	4*	7	7	-	-	-
Eczema	2	11	11	-	-	-
Minor da-mage skin	9*	9	9	-	-	-
Pruritus of skin	25*	25	25	-	-	-
Toothache	7*	7	7	-	-	-
Hemorrhoids (exacer-bation)	4*	18	18	-	-	-

	# cases	# recorded sleep	# of sleep in which dreams			
			were clear	were forgotten	unknown answer	not
Conjunctivitis and other eye disorders	5*	15	15 -		-	-
Closed fracture of thetibia	1	14	14 -		-	-
Osteomyelitis tibia	1	3	3 -		-	-
Polyarthritis	1	4	4 -		-	-
Endarteritis	2	6	5 -		-	-
Nocturnal enuresis	5	23	22	1 -		-
Total	353	1925	1642	97	121	65
% of sleep number	-	-	85,3	5	6,3	3,4

These diseases were noted several times in the same individual, and sometimes in people suffering from other illnesses.

Different diseases of the body clearly affect dreams. This has already been noted by Hippocrates[4] and Galen[5]. In our research we noted the change in the nature of dreams for some diseases and we try to define their diagnostic critical value. Subsequently, a number of researchers (Tsaturov, 1935 Anokhin, 1945; Lhermitte, 1948 Makarov,1951 Wolpert, 1951, 1966, IG Bespalko, 1955; Jost et al. 1955; Kparr 1956; De Benedetto, 1956; Whitty a. Lewin,1959; Ward et al., 19b1 etc) noted some changes in dreams for some nervous, mental and somatic diseases.

If even a mild paresthesia appears constantly in dreams it is easy to imagine more significant disorders occurring in the body will have an impact on the dreaming.

If doctors constantly monitor dreaming of patients throughout a disease and the period of recovery, they would often, if not always, notice changes in the nature of dreams.

On this issue, we have accumulated a lot of material collected in 247 people suffering from many different diseases. From minor cutaneous diseases, to mental disorders, and very serious, such as tumors of the brain. In all, 247 patients recorded 1642 dreams in different conditions: in hospitals, clinics and at home. Some of these dreams are recorded by the patients themselves, but the majority of dreams was recorded by us. This total includes the

---

[4] Hippocrates, a physician living in the fifth century BCE, believed that there were three different kinds of dreams. He said that some dreams could tell us about our ailments and possibly even provide a way to diagnose mystery illnesses. Other dreams were "revealing" - they provided information about something we were concerned about. Also, like the writers of the Gilgamesh story, Hippocrates believed that there were prophetic dreams. Hippocrates wrote a medical dream book: On Regimen 4 claims to offer a true account of the dreams that foretell physiological events. This account of such natural dreams combines explanation of their causes with suggested interpretations of their meaning.

[5] Galen, a physician who lived about four centuries after Hippocrates, took the idea of diagnostic dreams one step further. He used his patients' dreams to help him decide how to treat them - even to the point of performing surgery based on dream "recommendations"

dreams, which we will discuss: certain skin diseases, diseases of the throat, teeth, eyes, heart and abdominal organs.

- The character of the dream changes as the illness progresses. The attributes of the dream change at different stages of the illness, depending on:
  the severity of the disease,
- localization of the process, the development of the disease,
- the features that occur in the deterioration of various organs and systems.

However, when we analyze the changes in the dreams the most persistent ones are associated with features of disease and the characteristics of each person.

A latent disease will, in a vast majority of cases, increase the frequency and occurrence of dreams for one night and in developing long-term illnesses- for many nights.

In our total sample 54-76% remembered a dream, but the percentage of patients that remember a dream reached a level as high as the 85 - 100%.

Another common feature is that when a disease emerges, a patient will have dreams with a very unpleasant and even nightmarish character, with a predominance of troublesome visual scenes: war, fire, injury or other damage to different parts of the body, usually sick, blood, flesh, dead, grave, dirt, dirty water, poor-food products, the rise of the mountain or fall into the abyss, the hospital, pharmacy, doctors, medicines, etc. With almost constant displeasing thoughts, as well as feelings of anxiety, sadness, and fear.

For example, In the total number of 1642 dreams 1478, about 90%, are clearly undesirable in character: all of them had objectionable visual images, scenes, and feelings. In only 54 dreams an actual feeling of pain was noted, and that is not always clear. 1445 dreams (about 88%) had thoughts related to the content of the visual scene. In only 130 dreams there where no unpleasant elements associated with disease. These dreams were either in lung diseases or in the stage of recovery from serious illnesses.

Frequently the appearance of unpleasant dreams and dreams in general in various diseases were similar; it does not always depend on the severity of the disease.

In severe diseases such as a brain tumor, there were relatively fewer dreams remembered than in less ferocious diseases. However, this was due to some features of higher nervous activity in these diseases, which we shall discuss below. A common aspect of dreams that signal a disease is that they have disagreeable visual images and scenes connected with the peculiarities of the disease. These visual scenes were the most early signs of a disease, and they rather accurately reflected the location of the disease, the nature of the sensations and the functional features of the affected organs and body parts.

For example, in a case of gastritis there was a feeling of nausea with visual scenes of eating spoiled food and vomiting. During the sensation of pain in the stomach area there where visual scenes of wounds and injuries associated with bowel movements and the ache portrayed the intestinal region.

In lung disease, there were scenes of violation of breathing; in diseases of the genital organs there were unpleasant sex dreams; with heart disease there were nightmares and feelings associated with the heart area; with diseases of the muscles, legs, arms and torso- dream-scenes appear with traffic unable to move; with nervous and mental diseases dreams appear in terms of neuropsychiatric abuse etc.

The difference between the dreams of a sick person and unpleasant dreams in general, often due to day-residue like family or work problems; or even being uncomfortable during the night, while dreaming - too much light or noise, being too hot or uncomfortable in bed; is that the nightmares of sick persons:

1. Usually continue all night long and even during the whole illness period.

2. Have no direct link with exterior perturbing facts acting before or during the night.
3. The disturbing subjects of the nightmares (bad characters) are repeated in the same way in many dreams and have a link with a precise part or function of the organism; breathing, digestion, cardiac…

It is interesting to note that often unpleasant dreams appeared before other overt clinical symptoms of the disease. This is sometimes noticed by the patients, but many do not pay attention to dreams or forget them (but sometimes recall them after questioning). Only after questioning we were able to establish that the majority of patients suffered from sleeplessness and changing dream content before symptoms of the disease became obvious. This phenomenon has been observed in some disorders for a long time and attracted the attention of doctors (Hippocrates, Galen, Astvatsaturov, J. Lehr, Anokhin, Mayorov, etc.). Some doctors have used dreams for its diagnostic abilities and indeed, in dreams the onset of a disease can be found sometimes before other known symptoms are visible.

The initial changes in dreams may be different in various maladies. Dreams may be different, depending on the localization, process, the nature of disorders in the organism, and other features of the disease. To see this, we need to consider a change of dreams in various illnesses. Dreams from people suffering from different afflictions of the skin show that at the beginning of the disease dreams occur with resistant and visual images: scenes linked to the diseased portion of the skin showing its damage, and skin discoloration, dirt and other very unpleasant images. Sometimes these changes occurred earlier in dreams then the explicit manifestations of the disease. Here are a few examples:

K. medical student saw in the night of 6 7/XII 1937 in a dream that: "... *her roommate put a lighted cigarette to the rear right wrist, she pulled her hand back.*"

When she woke up, precisely in this place on the right hand she felt a burning sensation and slight pain, and during the day an inflammation in this place developed. This dream accurately reflects the location and nature of the irritation.

In the night of 12/VII 1945, a patient saw a dream in which "... *in the right temple area a swelling appeared in black, and then out of it pus flowed towards the right eye. There was a disturbing idea that if the pus gets into the eye, it can cause blindness.*"

While waking up there was a pain and slight swelling in the right temple, the exact spot that was noted in the dream, the pain was spreading towards the right eye.

In the afternoon of the same day the swelling at this site (the right temple of the head) increased, and then formed a small boil. Before this night there had been no sensation of pain in this region of the face.

Patient N. said that in the night of 17 XI/1956, he had a dream, wherein: "*. .. in his right thumb he felt pain, running from the nail to the base of one of the metacarpal bones. Then, on this place the thorn was filled with pus, it was painful*".

When he woke up, he felt pain in the nail of the thumb of his right hand in the exact place as the dream, spreading to the base of one of the metacarpal bones, just like in the dream. During the day, the pain in the finger increased, and there was a continuous feeling of agony.

In our sample 9 of these dreams in which severe skin disease appeared earlier than in reality, and the visual scenes correctly pointed out the place of injury and expressed the **specific** character of **the** sensations. But the subsequent nights, when diseases developed and caused general changes in body measurements, such as fever, worsening of the general condition of the body, the character of dreams changed significantly. Unpleasant scenery begins to dominate and nightmares occur.

Dreamer C. experienced in the evening 19/VI 1957 a strong pain in the end phalanx of the thumb of the left hand. His overall health diminished. Sleep in this night was anxious, C often woke up, had an abundance of unpleasant dreams, a nightmare, and all the dreams where connected with unpleasant scenes of the phalanx of his left hand. In one dream:

> *"He participated in the war. He was surrounded by soldiers, three of them with knives."*

C. tried to defend himself, but he wore a large helmet so he was not able see. One of the soldiers wounded with a knife the phalanx of his left hand, he saw blood and woke up in fear. There was a strong pain in the terminal phalanx the left hand, and his heart ached a bit. An inflammation of the finger occurred the next day, the pain diminished and dreams were more calm, however, persistent optic scenes with the left hand held on a few more nights, until recovery. Then dreams became more calm, however, persistent scenes with the left hand held on a few more nights, until recovery.

When we observe the dreams about skin diseases where recovery occurs, the dreams become normal, follow human nature and the displays of illness disappear. When the illness disappears, in dreams there are ideas and words connected with the disease, but they are not so vivid. Significant changes in the entire background of a dream, with a predominance of unpleasant **sensations** — boredom, fear and horror, usually coincided with general changes in the body: increasing temperature, malaise, anxiety and other disorders, but once passed, and these are common background symptoms in dreams that change for the better. Considering the frequency of the elements encountered in dreams, we observed the appearance of different injuries of the skin and subcutaneous tissue in dreams (Table 17):

table 17

Diseases	# of cases	# of cases falling asleep	# of dreams	# of dreams in which at least one of the following occurred										
				scenes linked with the experience of participants getting burned	thoughts & words	sensation of						Background of dreams		
						pain	pressure	cold	hot	itching	total	pleasant	unpleasant	nightmares
Furuncle	4	9	9	9	7	2	1	-	1	-	4	-	5	4
Burn	4	7	7	7	7	1	-	-	1	-	2	-	3	4
Eczema	2	11	10	10	10	1	-	-	-	-	1	-	7	3
Minor injuries	9	9	9	9	6	1	-	-	-	-	1	-	8	1
Permanent Itching	12	25	25	25	21	-	-	-	-	2	2	-	22	3
Withlow (infection)	3	10	10	10	8	3	1	-	1	-	5	-	6	4
Total	34	71	70	70	59	8	2	—	3	2	15	—	51	19

Table 17 shows that scenes related to affected areas of the skin occur in almost all dreams: 59 of the 70, about 84 %. In only 15 dreams there where different sensations observed: pain, heat, pressure, and itching. These feelings emerge in vivo in the recorded dreams of the more severe diseases, there are considerable local variations. The visual images and scenes change with the intensity of the pathological process.

If there is less damage to diseased places on the skin, the scenes in dreams often were more calm in nature. Conversely, when a disease is intense, scenes are unpleasant and often express injuries and damaged locations. The content of those dreams has a bad character, and express relatively local and general symptoms. In those cases a dreamer often experiences a nightmare. Dreams often occur before the outbreak of an illness. Sensations of heat and pressure, pain, unpleasant feelings (anxiety, fear) are only experienced as the situation gets worse. The dream-content will persistently show recurring troublesome scenes relating to the place of the disease.

Sometimes the first manifestation of the illness is observed in dreams. In nine cases of people with throat infection, six stated that the night before the outbreak of the disease they had unpleasant dreams associated with damage and throats. In nine cases of patients suffering from teeth problems, five had dreams with persistent negative visual content, associated with tooth decay.

> Another example: K. considers herself healthy on 6/IV 1955, but on the night of 6-7/IV, she had a dream in which *"she contracted diphtheria (a bacterial infection in the upper respiratory tract), clearly saw the pharynx, covered by a membrane of diphtheria."*

Her son who contracted diphtheria as well, also had this dream. He awakened with fear and anxiety. In reality, the sore throat was located on the right of the cervix, and the dreamer experienced palpitations. During the day K. developed lacunar tonsillitis[6].

In another dream on the night of 4/II 1950, a middle-aged woman saw that:

> "*Hooligans chased her, one of them grabbed her by the throat and began to squint it, she gasped.*"

When she woke up she had a sore throat, and difficulty breathing. And during the day she developed tonsillitis[7], although the day before she considered herself healthy.

With the development of angina and the manifestation of the common symptoms: fever, headache, pain and malaise dreams dramatically changed, they transformed and had unpleasant scenes (war, fight, nightmare, fear, suffocation), and scenes associated with the throat (pulling bloodstained ribbons from throat, discharging bloody slime and so on).

There is a definite relationship between dream content and the severity of the disease. If a disease worsens dreams will show especially unpleasant visual scenes, nightmares and fear. In total we recorded 26 dreams of patients suffering from an infection of the throat, in which patients remember scenes, this data is connected with the disease of the throat.

Problems in the mouth appear in dreams as visual scenes associated with an infected tooth. For example F.'s dream on the night of 15/V In 1958:

> "*The lower right molar tooth fell down, and the dreamer clearly saw three ruined roots in yellowish black.* »

Upon awakening he felt a slight pain in this tooth and during the day it increased and he had to go to the doctor. In this case, symptoms first appeared in the dream.

----

[6] Inflammation of the mucous membrane lining the tonsillar crypts.

[7] Inflammation of the tonsils most commonly caused by viral or bacterial infection.

Dreams also accurately reflect inflammatory diseases of the feet. For example, an elderly man N., suffering from chronic osteomyelitis of the right tibia, has dreams were he often sees his right shin in various unpleasant situations (operations, pollution).

In 1944, a patient got an enclosed bone fracture in the right tibia, in the lower and upper thirds. Being in hospital for four months, he had the opportunity to observe and note his dreams. He clearly noticed the relation between the nature of his dreams and the severity of his condition. During the first days of illness, when the body increases its temperature and insomnia occurs, dreams are very vivid, with colorful visual scenes. The dreams have offensive content from military life and always the right shin is involved. The right shin appears in dreams wounded, suppurated (pus formation), gangrene formation, operated and with **other** damage done to it. Gradually, after three weeks the pain decreased and the dreams took on a more calm nature; nightmares and fearful dreams became rare. The dreams were recorded in four months, but during the recovery they began to disappear.

We recorded 27 dreams from five patients with various diseases of the feet. The dreams contained undesirable images, based on the accurate location of the lesion.

Similarly, in patients with gonorrhea, the place of the disease was present in visual scenes of the dreams. In these diseases the content varied with the healing process: healing occurred when all the images related to the disorder disappear.

We recorded 17 dreams from 4 men with gonorrhea, and in all cases we observed distasteful imagery involving the genitals. Those dreams often contained sex scenes:

"... his penis was stuffed with gonococci (the bacteria that causes gonorrhea), which are then entered into the blood. »

And in these diseases as well, recovery varied with the content of dreams, the disappearance of the troublesome scenes was related to recovery.

In dreams concerning hemorrhoid the site of the disease is very precisely manifested. In four patients, we recorded 18 dreams, and in all case troublesome imagery related to the anus.

Dreams, resulting from internal disease are not so clear, but they always reveal a reflection of the localization of the disease, the nature of the process and the emotion of the dreamer. So, for example, diseases of the gastrointestinal tract are manifested in dreams as **troubling** mental images related to diet, digestion, or damage in the defecation process. During diseases of the stomach dreams often display sea — or spoiled food. If the bowel appears in dreams, visual scenes are associated with the act of defecation.

In more severe diseases in which the patients suffered pain, dreams contained scenes expressing the corresponding damaged region of the gastrointestinal tract by: wounding, surgery, and images containing tumors and damage.

Altogether, in 102 dreams concerning various gastrointestinal diseases[8] 100 cases had these scenes, 91 had a dream and observed thoughts related with content of visual scenes, and only 18 had dream sensations of pain, nausea and heaviness in the abdomen.

Acute and slowly beginning gastrointestinal disease sometimes manifests first in dreams. These manifestations are initially expressed in frequently repeated dreams with scenes associated with food not appropriate for consumption like raw fish, rotten meat, raw potato food with worms, and also defecation.

For example, on 13/VIII 1950 R., a middle-aged woman had, during the last month, often unpleasant dreams associated with eating raw fish. In waking life the gastrointestinal tract, gave some problems resulting in a a small appetite and other symptoms. When R. was examined, the doctor concluded she suffers from

---

[8] Gastrointestinal diseases refer to diseases involving the gastrointestinal tract, namely the esophagus, stomach, small intestine, large intestine and rectum, and the accessory organs of digestions, the liver, gallbladder, and pancreas.

night gastritis. In acute gastritis and enterocolitis9 changes in dreams occur more often at the beginning of the disease but become more clear while it develops. On the night before onset of the disease, dreams usually change content. Here is an example: In the second half of the day the dreamer ate a sausage of suspicious quality at a buffet. In the night these two dreams occurred:

> « ... I fished in yellowish-brown water. The fish I caught lies on the shore and quickly spoils, getting flabby and unpleasant to look at. The riverbank shows fecal contamination, this must contribute to the pollution and the damage of the fish. Now we can not consume the fish, because you can get sick."

The dreamer fell asleep again and had a nightmare in which:

> "... He was attacked by hooligans, escaped from them and went into hiding in the restroom of the Stadium in Leningrad. The restroom was dirty, with feces, urine and vomit everywhere. The dreamer had a feeling of disgust, and felt sick. The hooligans caught up with him in the restroom, and kicked him in the stomach."

The dreamer woke up in fear — experiencing abdominal pain, nausea, and had to go to the restroom, were immediately acute enterocolitis manifested.

The feelings in the abdomen at awakening did not appear in the dream whereas the imagery was clearly unpleasant and associated with the corresponding sensations of gastrointestinal disorders.

Therefore its emergence in the dream scene can be regarded as the first manifestation of the disease. Falling asleep during the

---

9 Enterocolitis or coloenteritis is an inflammation of the digestive tract, involving enteritis of the small intestine and colitis of the colon. It may be caused by various infections, with bacteria, viruses, fungi, parasites, or other causes. (source *Wikipedia*)

second phase of the disease, it manifested itself through certain scenes associated with it.

An elderly man suffering from chronic gastritis and colitis said that in dreams:

"...He often eats earthworms and raw bloody meat, and out of his mouth and stomach comes unpleasant mucous tape and pieces of slime."

As already mentioned, different sensations during the gastrointestinal transport-act and its appearance in dreams, as a rule, states that scientifically the awake state and the dream state are connected. Feeling sick, often causes scenes, expressing different parts of stomach-damage, injury, operations, ulceration, abscesses, etc., sometimes fairly accurate in its localization.

For example, a physician N. suffering from chronic colitis, reported that on the night of 17/II 1959:

"... On his stomach, a little to the left of the navel appeared a a red painful lump on the skin, this boil is discovered, there was an image of pus and blood and it formed a fistula[10], the abdominal wall covered with pus and feces".

Upon awakening there was pain along the large intestine, but more in the navel area, on the left. We recorded 54 dreams with severe abdominal pain.

Surgeon L. eats fish that is not quite fresh on the evening of 12 / II 1958. In the night of 12 to 13/II 1958 he dreams:

"... While visiting patients, I become ill and vomit".

L. sees the stomach, his intestines and has anxious thoughts. Upon awakening he feels a strong abdominal pain, vomits and has discomfort in throat and mouth.

---

[10] A fistula is an abnormal connection between an organ, vessel, or intestine and another structure. Fistulas are usually caused by injury or surgery, but they can also result from an infection or inflammation. Fistulas are generally a disease condition, but they may be surgically created for therapeutic reasons.

A feeling of nausea is often associated with such visual scenes: spoiled food, faeces, the act of vomiting, worms, and sometimes a feeling of disgust and nausea.

Dreamer H, sick from chronic gastritis in the night from 5 to 6/11, 1954 in a dream

"... took a bath contaminated with faeces, it was sick, I tried to vomit, I persistently sought a place to do that, but everywhere there are people."

Upon awakening the dreamer experiences heaviness in the abdomen and a feeling of nausea. In a dream on the night of 23 to 24/VII 1954:

"... ate fish of a pale yellow color, half-baked, flabby consistency and some sauce with an unpleasantly dirty yellow hue. And in the dream I was disgusted by this food, but continued to eat."

And when he woke he felt a heaviness in the stomach and mild nausea.

These kind of scenes are recorded in 43 dreams during gastrointestinal tract infections and nausea. On awaking, 21 of them attempted to- or vomited during the night.

If we look at the way dream content develops during the process of gastrointestinal disorders, it is easy to see that it varies depending on the expression of symptoms and the nature of the disorder, while all this is most clearly manifested in visual scenes. The early symptoms usually begin to appear in relevant visual scenes, in the midst of the disease they take on a more unpleasant nature, and in the recovery phase they disappear. Often in these dreams unpleasant thoughts occur, related to the content of the visual scenes, sometimes unpleasant emotions like anxiety and very rarely the sensation of pain, nausea, etc.

As already mentioned, for some gastrointestinal disorders, unpleasant dreams sometimes express day residue that is related to the disease. Out of four cases of dysentery[11], for example three dreams showed changing dream content related to the disease the night before, and one dream two days before the explicit manifestation of the disease.

One of the cases is a paramedic who had, two nights before, dreams about acute dysentery on the morning of 8/VII 1950. On the night of 7/VII he has a dream about:

"... an apartment contaminated with feces, dirty water yellowish-gray" and on the night of 8/VII he had a dream of a nightmarish character: *"He participated in the war against the Germans. He was hunted by a group of Germans, he escaped from them hiding in the restroom, which was very contaminated with feces. He tried to urinate, but at this time the Germans found him. The Germans started to beat him on the butt and on the head. The second bayonet thrusts in the lower part of the stomach and he feels the pain."*

In fear the paramedic woke up in severe pain. His lower abdomen was aching, like in his dream, and he had a headache. The patient was sent to hospital, where he was diagnosed with acute dysentery. In acute dysentery, most dreams are especially unpleasant nightmares, with always unpleasant visual scenes related to the abdomen, especially the bowel.

In 1943 we saw an outbreak of infectious hepatitis (Botkin's disease). It was observed that sleep and dreams in a few patients changes the day before the onset of the disease, and becomes disturbing and superficial. There are unpleasant dreams about war, etc., there is a strong feeling of melancholy, fear, and anxiety. All

---

[11] Dysentery (formerly known as flux or the bloody flux) is an inflammatory disorder of the intestine, especially of the colon, that results in severe diarrhea containing blood and mucus in the feces with fever, abdominal pain, and rectal tenesmus (a feeling of incomplete defecation), caused by any kind of infection. It is a type of gastroenteritis.

this accompanied by unpleasant scenes concerning the right hypochondrium area[12], in the dream this place is cut and damaged.

In the other two cases of acute dysentery the unpleasant dream was the night before the manifestation of the disease, while in one case, the disease was revealed immediately after awakening, and at other times six hours after awakening.

Dreams about the exacerbation of chronic cholecystitis[13] are most often associated with with pain in the liver and gallbladder. In our sample we observed 24 such dreams. Three times we were able to record the dreams of people who became ill with acute appendicitis.

In the first case, a medical student dreamed: in the night at 2:01, she had a dream in which she was:

> "...buried in the ground, half-clothed people are throwing land on her abdomen and chest, her stomach feels the weight of the earth."

She woke up sweaty and scared, feeling bad, and two hours later developed acute appendicitis.

P. complains in the evening of 11/ VII 1951 about abdominal pain, in the right iliac region, but does not have obvious symptoms of appendicitis. In the night of 11-12/VII, a night of disturbed sleep, P. has a lot of dreams, all unpleasant. Visual scenes associated with the disease:

> "...being cut in the belly by doctors", "being wounded in the belly in the war". and "the belly itself broke up into ribbons and from a crashed intestine".

P. wakes up with severe abdominal pain, and is diagnosed with acute appendicitis.

_____

[12] The hypochondrium is the uppermost part of the abdomen, just below the chest or thoracic region.

[13] Inflammation of the gallbladder.

In the third case officer S. is healthy in the evening of 10/IV 1956. On 11/IV S. dreams:

"... was in the war, he was wounded in the stomach, clearly saw a large wound in the lower right half of the abdomen."

S. wakes up in fear. When the doctor examines him he is diagnosed with appendicitis.

When one looks at dream content in lung disease, unpleasant dreams are usually associated with the chest and the act of breathing, such as:

- resting on the earth,
- something massive is placed on the chest,
- getting through a narrow opening,
- wearing uncomfortable clothes that hindering breathing, and feeling suffocated,
- anxiety, fear.

We recorded twenty-two dreams of four people suffering from light sensitivity for pulmonary tuberculosis during a period of ill health. Most dreams had visually unpleasant scenes associated with this disease. Moreover, in all these cases there were anxious thoughts about the disease. In any long term disease, if the patient becomes aware of the diagnosis, disturbing dreams associated with thoughts about this disease begin to appear.

A medical student, who suffers from pulmonary tuberculosis, (compensated stage) about three years told me that in dreams she often sees herself sick, suffering from this disease and thinks about it. For example, on the night of 5/IX 1936, she has a dream in which:

"... She lays down naked on the damp wet ground, as she suffers from pulmonary tuberculosis. Then the earth beneath her sinks, as if forming shelved edges which then converge and squeeze the chest, making it hard to breathe."

In fear she wakes up and experiences troubled breathing, feeling bad, cold sweat. She **even** had the same type of dreams two months before detection of pulmonary tuberculosis.

The second patient, **with** pulmonary tuberculosis for two years with symptoms of exudative pleurisy[14] on the left, said that in the moments before getting ill she often has unpleasant dreams in which her chest is **heavy** (crashed by mountains, land) or she has to crawl through narrow openings that get stuck on the chest.

Similar dreams were for acute bronchitis, runny nose, and other conditions that impede breathing. We recorded 67 such dreams. An example: officer P. with acute catarrh of the respiratory tract in the night of 14/VII 1951 has a dream in which

> "... was in the war, he was attacked, piled on the ground and was being strangled. He felt suffocated."

P. woke up in fear breathing heavily, with a stuffy nose, and a headache.

During an acute bronchitis with a runny nose on the night of 8/X 1954 P. was dreaming:

> "I'm walking with comrades on a muddy road, it sucks. They quickly climbed a mountain, I also walk hard, I feel uncomfortable, my helmet feels like a whip all the time and prevents me from watching. Then we walk along the the river, the water is yellowish dirty, I am under extreme exposure, stumble and fall from the shore into the river, deeply immersed in water for a long time. I can not come up, start to choke, feel pain and think that now, perhaps, my life will end."

In fear P. woke up and sighed with relief, glad it was only a dream. He finds himself lying down on his face, nose, his head aches, a bruised feeling in his body, as if the chest is squeezed.

---

[14] Pleurisy, also known as pleuritis, is a condition that results from the swelling of the linings of the lungs and chest.

Scenes expressing the inability to emerge from the water for long time, were observed in 11 dreams and they were all observed in patients suffering from disease of the respiratory tract. Other health-related dreams also can show visually unpleasant scenes involving breathing and anxiety.

There are certain characteristics about various heart diseases. In the past, a number of authors noticed peculiar dreams during cardiopulmonary disorders[15].

For example, Astvatsaturov[16] (1935) believed that disturbing dreams with elements of fear of death, ending suddenly by awakening, indicate cardiovascular disease even in the absence of other objective data. Such dreams **were** described **by** Lhermitte (1948) and other researchers. Indeed, even minor changes of cardiac activity during sleep, such as palpitations, are often marked with nightmarish dreams, visual scenes and feelings of fear.

In patients with myocardial infarction and angina pectoris characteristic dreaming appeared: nightmarish scenes related to the area of the heart, and a strong sense **of** fear of death. P., 51 years, had suffered a myocardial infarction three months ago. A fortnight ago she buried her husband. She often has nightmares, especially disturbing her during times she is not feeling well. Here is one of them:

All day 12//V in 1954, she felt a little pain in the heart, and on the night of 13 /V she has the following dream:

"I was at the cemetery, sitting on my husbands' grave. Suddenly from the grave two bony hands rose up and grabbed me by the throat. One, hand dug into the heart area, I see-

---

[15] Cardiopulmonary diseases are the medical terms used to describe a diverse group of serious disorders that affect the heart and lungs.

[16] About dreams with heart disease, Astvatsaturov wrote: "And if patients have heart frightening dreams that in reality is often the case, then we have every reason to regard them as a product of the somatosensory (more viscero-) mental shift" i.e. body-visceral switch.

med to feel the bony fingers dig into my body, it became difficult to breathe, I was frightened, and wanted to cry "help", but was not able, to. I finally cried and woke up in fear." There was a strong heartbeat, pain in the heart and throat spasm.

An elderly doctor A., one day after an acute myocardial infarction, was in the hospital; in the evening of 25/V 1962, his health was satisfactory, he does not feel pain in the heart. On the night of 25 - 26 /V he had the following dream:

"In a native village he lies on his back on a stove in a stuffy cottage. Suddenly the house begins to break down and a fight starts. Thieves are fighting with him. He clearly feels fear. One of the gunmen climbs onto the roof of the hut and takes out logs of the ceiling beam. These fall on his heart area. It feels heavy, and the dreamer experiences an intense fear. He tries to wriggle out of the logs and wants to scream, - but is not able to, in the end the dreamer shouts and wakes up in fear."

Lying on his back, he felt warm all over, feeling a squeezing pain in the heart. **According to his neighbors he had uttered inarticulate sounds and shouted before waking.** In this dream, pain in the heart manifested in nightmares and, in particular, the impact of a beam in the region of the heart, feeling the heat - lying on the stove at the same time and on the back - a common position in bed for this patient. The fact that the patient has not been able to cry out, indicates difficulty in the speech centre of the brain.

Middle-aged officer P. who has been suffering from angina for three years said that on 14/III 1956 during daytime, he felt a slight pain in the heart area which held on all day, then increasing, then decreasing, in this state, he went to bed . In the night of 14 -15/III he had a dream in which he:

"...was in combat operations, he was attacked by several enemy soldiers and tried to retreat, but from all sides sol-

diers came closer and closer, one of them stabs him in the heart area with a bayonet. P. has a strong fear : 'this is the end of my life'."

When P. wakes up he feels pain in the heart and this fear persists some time after awakening.

We recorded fourteen dreams about angina and recent myocardial infarction (see Table 3), and in all cases, there are nightmarish scenes associated with the region of the heart, and a sense of fear. Thus, we see that for diseases with the potency to severely destroy various local and fractional bodily functions, dreams about **authorities** appear, related to the location of painful processes and functions, affected by the frequency and the nature of the body sensations.

The most distinct changes in dream content appear in diseases that affect the whole body or diseases that are acute.

So, for example, for influenza and acute respiratory catarrh dream content dramatically changed:

- Unpleasant and nightmarish imagery.
- War damage to various parts of body-and head: throat, nose, and other parts that are threatened by the disease.
- **Often** in such dreams were scenes related to mud, muddy water, raw meat, fish, lifting the mountain, coffins, corpses, hospitals, doctors, etc.
- Often an unpleasant thought, expressing doubt, failure, anxiety, dissatisfaction, a sense of longing, fear.

We recorded 77 such dreams, 36 of flu and 41 with acute respiratory catarrh tract. Sometimes a change in dream content appears before manifestation of overt symptoms of illness, usually the night before. In 18 cases there had not been a noticeable manifestation of the disease but in the night dreams acquired an unpleasant often nightmarish character and in the morning or during the day the disease developed.

A medical student on 31/I 1938 at home, his health is satisfactory. No specific incidents during the day. His sleep on the night of 31/I is restless, he often wakes up and is dreaming a lot, all dreams have unpleasant content. In one he was:

> *"...involved in the battle, it seems, with the Germans, beat him on the head, back, he tries to hide, crawling through the mud mixed with snow, then swims across a river with muddy, dirty water, his clothes floating on the back of his rifle, it all pulls him to the bottom. He begins to sink, and he, in panic, thinks he is going to drown. He feels a sense of doom."*

The student wakes up in fear with an unpleasant sensation throughout his body, especially in the back between the shoulder blades, fatigue, aches in his head, the notion that his night was full of fights and harassment. During noon the medical student developed acute catarrh of the upper respiratory tract.

A Doctor, 41, spent the day of 25/VIII 1950 doing his usual work in an institution. His state of health is satisfactory, and in the night of 25 - 26/VIII he falls asleep quickly, but then the dream become alarming, he often wakes up. There are many dreams with bad content.

In one dream he:

> *"gets in Kazakhstan. He rides in the mountains, very steep mountains that the horses can not climb. Traveling is difficult, as more rain falls, the feet **plastered** with dirt. He is very tired, restless thoughts appear like: will I be able to get out of these gray, mud-covered mountains?. Finally, he was able to limp into the village and finds himself with a familiar Kazakhs, who according to an old custom cut the throat of a sheep, pouring scarlet blood in **bright** spots. It gives him an unpleasant alarming feeling. He sees how the carcass from the sheep is cut; the meat, the fat, (red with yellow streaks of fat). Then he and others began to eat raw meat, there's a disturbing thought: do not get sick from eating this raw meat."*

In the next dream:

> "… *Since then the comrades in the service are involved in a clash with police. A lot of them, they are winning. He hides with his comrades in the forest,* **carries** *provisions, including large chunks of raw meat, on the back and on the head, it makes it difficult to walk at all.*"

The Doctor wakes - feeling unpleasant in the parietal part of the head, in the back, general malaise, similar to what happens in the initial stages of flu. During the day he developed a moderate flu. From the words of the doctor: if he has unpleasant dreams of an undefined character in which there is raw meat and especially if he eats it, it is usually associated with disease. In his case, raw meat is so firmly bound with illness that in the dream he remembers that eating raw meat is associated with disease, so he is trying not to eat. He told me that once during a catarrhal disease on the night 2/IV 1954 in a dream saw:

> "*meat and it was not even his, but then he remembered not to eat meat in a dream so he spit it out.*"

Many individuals have indicated that frequently before the disease with increasing body temperature in dreams there is visual exposure of body scenes like bathing in dirty water, or cleaning the bath. For example, P, on the night of 5/X 1959 was healthy, and on the night of 6/X dreams of:

> "*the bright summer sun bathed in blurry water, it seems to have been warm. When out of the water - had signs of yellowish dirty water, wiped his hands.*"

Upon awakening the entire body was covered with sweat, and the unpleasant feeling of soreness and heat throughout the body. In the afternoon of 6/X the dreamer gets the flu with high temperature. Patient N. 2/I 1959 reported that if he dreams about dirty water and dirt in it, then the next day, as a rule, he develops a cold.

Dreams at the height of a disease are visualized as particularly nasty, with a predominance of nightmares, fear. As the disease gets worse, the nightmare will become more pronounced. However, in any period of the disease one can notice certain changes in the dream associated with the more severe symptoms of the disease; for example, a severe headache will often be visualized in scenes associated with different changes to the head; injuries or diseases of the throat will be visualized by scenes associated with damage to the throat, while the heart with scenes expressing damage to it, etc. Such dreams were already cited, when we looked at the dreams of other diseases like headaches, throat, heart, and disorders of the gastrointestinal tract. How accurately various changes and sensations are reflected in dreams can be seen in the following dream.

During lung catarrhal state on the night of 31/VII 1954, among the many dreams one had the following content:

"I've been involved in a fight, we are moving on red horses alongside an avalanche in Kazakova. The faces of the Kazakov's are red, they have big red beards, their mouths are open, they scream and cry. Some of them carry lances, others are waving their swords. We are with few men, but we ask to shoot. The avalanche is bending, there is a feeling of fear and anxious thoughts. I find myself surrounded by a Kazakova who tries to stab me with a saber. He stabs into the left side of the neck, I feel at this point the edge and try to cover my face. The Kazakova unbuttons my shirt with the saber, the buttons and the tip of the saber are in the same place. I am trying to protect myself from the Cossack by raising my right hand and grip the sword, I feel the sharp blade in the palmar surface of the middle phalanges. There seems no way out, I feel a sense of impending doom. I wait with horror for a sword to thrust into my throat and kill me. I try to scream."

The dreamer woke up in fear with a sore throat, more on the left, on the same side an increased painful submandibular gland[17], a sharp tingling on the surface of the middle phalanges of the right hand, general weakness in the face, of a feeling of warmth, heaviness in the head and tinnitus.

In this dream, the presence of pain in the left submandibular gland and left area of the throat are reflected by the wounding checkers of this place, and even the pain reflects the sensation of sharp tingling in the hands of the dreamer when he covers his face.

The pain in the right second phalanges manifested while grasping the blade of the sword and feeling a sharp blade at the same time in these places.

The feeling of warmth in the face is reflected by the large number red bearded individuals. The feeling of fear and doom in the dream was probably a manifestation of common disorders in the body, and possibly a consequence of the visual nightmare scene. It is interesting to note that the various sensations in the head and face during sleep in dreams are often visualized by muzzled horses and other animals, and they will sometimes be disfigured. In the further course of the disease the scenes are modified accordingly and get better and return to normal for each person.

In 1947, in one of the hospitals we had to observe patients with typhoid and typhus, and we recorded thirteen dreams in five patients. Some people told me that already a few days before the onset of disease, they noticed changes in sleep and dreaming. For example, of the three patients we observed two typhoid patients who definitely pointed out that a week before the diseases they observed drowsiness, and sleep was troubled with unpleasant dreams, and we observed both typhus patients saying that for five or six days, dreams began to take an unpleasant and even horrible

---

[17] The paired submandibular glands or submaxillary glands are major salivary glands located beneath the floor of the mouth. They weigh about 15 grams and produce around 60–67% of the total volume of saliva.

character. In the midst of the disease in patients with abdominal typhus had nightmarish dreams, with anxiety, fear, and the dreams were particularly colorful. Patients say their dreams passed in a blur, the details are often forgotten when recorded.

A patient aged 36, on the third day of illness said that on the night of 21/IX 1947 she saw several dreams, all unpleasant, but she forgot many. In one:

> "*Its as if I am buried in the ground, thrown on the ground, my belly seems to feel the gravity of the earth, it was hard to breathe, I am afraid to die. Then I began to sink into the ground more and more, there was darkness, the darkness seems somewhat faded in this dreaming and I awake.*"

The dreams of typhus patients looked somewhat different, although they are also dominated by nightmarish scenes, anxiety and fear, but expressed in very colorful scenes. Patient D., 31, soldier in the Great Patriotic War of 1941 - 1945 on the second day of illness told me that in the night of 21/VII 1947 he had many dreams about military life, in one dream:

> "*He was attacked by enemy soldiers, very scary, they had big heads, big fiery red bulging eyes in their faces, red hair. He was surrounded, beaten with rifle butts to the head, he drove his sword in the head, saw the blood on the face of the enemy but was overpowered. He flees but is pursued, and is continuously beaten on the head.*"

Such frightening dreams continued throughout the disease, and then they took on a normal recovery character.

Dream content changes slow during the development of a common disease. For example, in hypertension dreaming was very bright, with a predominance of unpleasant visual scenes connected with the area of the heart, and always with a sense of anxiety and fear.

It is interesting to note that patients who suffer from hypertension indicate a change in sleep and dreams about 2-3 months in advance of the diagnosis.

Civil engineer G. in March 1959 said that since the end of 1958 he had a disturbed sleep pattern, and unpleasant dreams; most often associated with construction. For example, on the night of 21/111 in 1959 in the dream,

> "*he is up on a house being constructed based on his project. The house is built careless with cracks in the wall, and it even rocked; he had a feeling of anxiety that he may judged for such work. Then it all collapsed, it filled up, he was crushed by the head, chest, and had trouble breathing, there was a strong fear.*"

Upon awakening his head ached, he felt a little dizzy and sensed heaviness of the heart, and an awareness of fear.

In five hypertensive patients we recorded 12 such dreams. Thus, we see that under a variety of diseases in the dreaming there always appeared scenes where things were connected with the peculiarities of the disease. If you take all 454 dreams, noted in a variety of diseases: internal organs, skin, muscles (Table. 3), except for diseases of the brain, we notice that visual scenes, associated with the nature of disease, met, in all cases, unpleasant thoughts. 340 dreams of illness are associated with sensations of pain, heat, heat, nausea and dizziness.

Let us now try to trace the characteristics in dreams as a result from brain disease. To do this, we first consider the dream appearing in functional disorders of the brain. During many years, we have collected a lot of dream persons suffering neuroses, only 19 people recorded 573 dreams. Our observations relate mainly to two types of neurosis: hysteria and neurasthenia. The dreams of neurotic patients without significant violations in the brain functions are interesting in the sense that their dreams show the most distinct changes in the body and the influence of the external environment. If the dreams of patients using medications are very sensitive to stimuli administered during sleep and the slightest changes to the body; then, in a healthy person it would have no effect on dreams, or it would go unnoticed. In this respect. studying persons suffering from neurosis can give us deep insights into the interaction of the brain with its external and internal environment during sleep. And I must say that the most interesting general patterns of brain activity during sleep are seen in the seams of people suffering from neurosis.

Our observations consider neurasthenia[18] and hysteria. The sample consisted of 19 patients we observed a long time, 3 people were treated on an outpatient basis and only 6 are periodically placed in stationary hospitals. Five of them did not stop going to their job, three housewives and one pensioner did not work. You can see the distribution of the sample in Table 18. It should be noted that, besides those identified in separate dream groups of persons suffering from neurosis, we have a large number of dreams recorded in functional disorders of the nervous system. These dreams are considered in relation to major diseases or physically disabled.

Table 18: Neuroses and dreams

Disease	Men	Women	Sum	Number of dreams
Neurastenia	10	1	11	487
Hysteria	1	7	8	86
Sum	11	8	19	573

[18] Neurasthenia is a term that was first used at least as early as 1829 to label a mechanical weakness of the *actual nerves*, rather than the more metaphorical "nerves" referred to by George Miller Beard later. As a psychopathological term, neurasthenia was used by Beard in 1869 to denote a condition with symptoms of fatigue, anxiety, headache, neuralgia and depressed mood (source Wikipedia

Disorders of the nervous system are known to disrupt sleep, sleep becomes superficial, shallow, with lots of dreams. Often the first symptom is an abundance of dreams. In our sample of 638 observed cases there clearly where 573 (about 90% of 628), 24 cases where, but are blocked and in 23 cases the answer was uncertain. Only 18 cases (about 3%) specifically claimed not to experience/ remember dreams.

Not only the frequency, also the content of dreams changed. Even the most minor functional disorders of the nervous system, those disorders that are not visible or attract attention are reflected in dreams. Persistent dreams with scenes of an unpleasant nature like: failure in school or work, backlog of trains, war fighting, fire, raging sea, drowning people, nasty sex scenes, etc. with elements of frustration, anguish and fear. The content of these elements is personal and depends upon the education, occupation, living conditions, the characteristics of the organism and the disorder. But in general those dreams have unpleasant emotional and visual content. And indeed, in all cases when the body experiences even minor violations due to the disease, in dreams scenes and emotional elements associated with the threat will appear. The heavier the violation to health, the more pronounced changes in dreams occur. Such phenomenon has been noted almost always, *except when the nervous systems is very weak and is unable to sufficiently respond adequate to stimuli that happen during a long, very severe disease* (tumor in the head, a malignant brain cachexia[19], sepsis; whole body inflammation etc.).

Let's go back to dreams with nervousness. We have already said that the main content of their predominance had unpleasant content. And indeed, if you view all 573 dreams recorded by us in

---

[19] Cachexia is a common complication of cancer (and some other medical conditions such as HIV/AIDS) that is felt to be responsible for 20% of cancer deaths in the United States. That said, it is rarely diagnosed until it has been present for a lengthy period of time.

neurotic patients it appears that 561 dreams contained unpleasant scenes and feelings, only 12 did not have such unpleasant scenes and visual sensations.

Unpleasant thoughts, related to dream content were recorded in 505 dreams (90%), feeling sad, anxious, fear in 443 dreams (79%).

We observed 19 patients with neuroses who stated that sleep and dreams changed before the explicit manifestation of the disease. At night they were more likely to wake up, could not sleep because of light or noise, and the slightest external stimulation caused awakening. In some this detection happened 1-2 weeks earlier then the manifestation of the disease, in others a few months earlier.

For example C, 19 years who prepares for admission to an Institute on May 1940 said that since February, he used to wake up sometimes very early, at 4-5 o'clock in the morning, with a sense of longing. In his dreams there are many troubles, he fails the examination for the institute and he was left alone. Gradually the content of the dreams is becoming more unpleasant. He frequently dreams:

> "*he was drowning in a raging sea, on a shipwreck, and falls into a dirty deep river.*"

He awakes with a constant feeling of anxiety and fear. There where many unpleasant dreams. At the end of April 1940 this young man displayed signs of neurasthenia.

Another asthenic type[20], second course medical student at the N. Institute, who had a neurotic breakdown during the winter exams, said in February 1937 that the unpleasant dreams and nightmares

---

[20] Astenic type: a theory that was developed by the German psychiatrist Ernst Kretschmer. In his book *Physique and Character*, first published in 1921, he wrote that among his patients a frail, rather weak (asthenic) body build as well as a muscular (athletic) physique were frequently characteristic of schizophrenic patients, while a short, rotund (pyknic) build was often found among manic-depressive.

started to emerge 1,5 - 2 months before the explicit manifestations of neurasthenia.

When we trace back how early patients noted a change in dreams if the disease is neurosis, we obtain the following information:
In a total of 19 patients:

- 2 had experienced a changing character of their dreams, 1-2 weeks before the explicit manifestation of the disease;
- 5 patients - 3-4 weeks,
- 6 patients - 6 weeks ,
- 4 patients - 2 months,
- 2 patients - 3 months or more.

Different people had neurasthenic symptoms manifested in dreams, in different forms ranging from minor unpleasant scenes or dissatisfaction to nightmares and strongly expressed feelings of fear.

For many years we continue to observe the dreams of N., who has a weak nervous system, and often suffers breakdowns by type of neurasthenia. N was constantly troubled by his inability to perform, he was inferior in his work and did not get promoted. Over the years N. often has the same type of dream about the same event, namely not being up to the job, being last in the race, missing the train, and these scenes are continuously combined with a sense of dissatisfaction. Here are a few of his dreams:

In the night of 1 to 2 /II 1938 has a dream in which:

"... N. is at home, his father and younger brother are with him. They make hay, but all the time N. works slower as his father and brother, and is not able to perform easy job. N. wakes with a sense of incompleteness."

In the morning N. wakes up feeling bad, with a heavy head. On 14/II 1938 his health is satisfactory, but he has some stomach trouble. In a dream on the night of 15/II:

"is on a steep hill with fellow students, but can not climb. Then he sees a river with steep banks. The river is whitish - muddy. His fellow students jump in the water. He also gets in the water and tries to cross the river, by swimming, but can not keep up with his comrades. Then he goes on a narrow board, over the river, on this slippery plank, and gets to the opposite shore with great difficulty. His comrades laugh at him. He woke up with a sense of dissatisfaction.

While waking up he is in a bad mood, and has a sore stomach. This dream reflected the general state of weakness and feeling of the gastrointestinal tract. Later on, when he was already a doctor, but could not make a living the way he wanted, N. began to notice that he fell more and more and more behind on his comrades in the service. He often had repeated dreams in which he missed the train. At this time in waking life he had often traveled by rail, but there where dreams during this period of life when he did not feel well. We also give one example of these dreams. During the holidays, on the night of 21/V 1950 he had a dream in which:

"... He was going to leave the place of service. At the station, he went to the restroom, and wants to empty his bladder, commits this act, but can not finish. He hears a beep from the locomotive, hurries to the train, but the train has already departed from the platform. 'Late again' he thinks, and catches up with a lot of difficulty and climbs in the last car."

Upon awakening he feels an unpleasant sensation, a filled bladder, and a sound coming from the kitchen: the ringing of utensils. This dream reflected his constantly disturbing thought of inferiority, irritation of the bladder and possibly affected by sound stimuli — the clatter of dishes in the kitchen. In his stories, for many years, he had a dream in which his train departs, but he still can catch up. In 1955 he came up to me and said:

"Today, I was quite late, could not catch the train and left the station almost in my underwear."

His mistakes, that occurred frequently in his dreams manifested his thoughts, doubts and concerns about his weakness. These dreams occur quite often at moments of nervous breakdowns. On several occasions when he had a period of developing a state of neurasthenia, his dreams as a rule, were very very unpleasant in nature, filled with nightmares, depression, anxiety and fear. So, in the afternoon 5/X 1952, his state of health is poor, with feelings of fatigue and weakness. In the night of 5 - 6/X:

> *"N., his older brother and an employee found the body of a giant man, who died many years ago. The corpse is sewn into the skin, lying on his stomach. One of his colleagues reveals that the skin underneath the skeleton is brown, covered by muscles. The dead man revives and grabs him by the right side in his stomach an he feels these places touched."*

N. wakes up in fear. Lying on his back, he feels pain in the liver and stomach, he still feels anxiety some time after waking.

In the following nights his dreams where dominated by unpleasant scenes, but in the morning dreams became better and better and before waking the nightmares disappeared. It is interesting to note that in fatigue, whether caused by neuritis, as well as mental and physical fatigue, nightmares often mark the beginning of the night, and the dreams during the morning hours are more calm in nature.

Almost all neurotic dreams are dominated by troubles or a nightmarish content (about 97.5% of the total number of dreams). Our sample has made clear that the nuisance was different for every patient.

C., a young impotent man due to neuritis and weakness, which troubled him most, dreamed about unpleasant visual scenes of a sexual nature. P., who had functional cardiac activities with severe pain in connection with neurasthenia, would have nightmarish dreams, with scenes of the affected area and a strong fear of having a heart disease. The third student, K, who simultaneously suffered from neurasthenia and colitis[21], often encountered in dreams nightmarish scenes with an impaired gastrointestinal tract. Very often in the dreams there was unpleasant imagery related to head injury, uncomfortable outfits, uncomfortable position, changing forms, and it always coincided with a headache during sleep.

The dreams of neurotics are usually associated in an unpleasant way to a situation that contributed to the emergence of neurasthenia. A second-year medical student N., diagnosed with neurasthenia, during the exam preparation for normal anatomy, often had dreams associated with discipline:

"N is in an anatomical theater, preparing corpses, but the skeletons attack him" etc.

This student said that in time, many years after he was cured from neurasthenia, often he is dissecting these bodies in dreams when he suffers from fatigue. For the dreamer those dreams are a warning sign against fatigue so that he will reduce his mental workload.

Apart from having bright visual scenes, usually with large pictures expressed in different colors, the dreams of neurotics are more definitely related to everyday assumptions about the limits of their own possibilities than dreams of healthy people. A neurotic, afraid of public speaking in waking life, has dreams of exultant scenes in which he is a presenter for a large audience, speaking with confidence and logic, outlining his thoughts.

One of the patients, said that in dreams he almost always has meetings where he acts much more logical and calmer than in reality. However, at the same time there are dreams in which he was *"Studying in secondary school and is not prepared for the exams."* although he has higher education. Such dreams were also observed in other neurotics. But most often the thoughts in dreams from neurotics are unpleasant in nature, expressing doubts, anxi-

---

[21] Inflammation of the inner lining of the colon.

ety, failure, usually connecting nightmarish visual scenes with the life situation, that promoted the neurosis. The elements in neurotics's dreams more often, than in healthy people, display:

- hard sounds,
- heat,
- cold,
- touch,
- pressure,
- pain, etc.

Neurotics are very sensitive to external and internal irritations. The slightest change, occurring in the body, and external stimuli operating during sleep, cause extremely vivid dreams in those patients. Feeling sick, N. suffering moderate neurasthenia in the night of 19 and 20/IX 1953, had a dream in which:

"... He was haunted by an old bloated man and a woman who looked like witches. At last they caught up with him, he had to belch, squeezing his chest with his hands, his breathing became difficult because of the witch on his back. He fled in fear, trying to throw off this unpleasant burden, but could not, his chest felt increasingly squeezed. His breathing became difficult, he felt like he was going to die from suffocation. He tries to scream but is not able to, and when he finally is able to shout he wakes up in fear."

The dreamer woke up lying on his back, with an unpleasant feeling of pressure in the chest, difficulty in breathing, experiencing total fear.

A middle-aged man suffering from neurasthenia, had a lot of troubled dreams on the night of 19/I 195, in one:

"... I was in some mysterious country. I experienced a feeling of anxiety and crawl into the narrow opening under the fence. I got stuck at the chest. Behind me there were two strange figures with round heads, shiny red eyes, and long

red tentacles instead of hands, like bent rings. I felt fear, trying to get out from under the fence, scraping the ground with my fingers and even felt the hardness of it."

Upon awakening the dreamer felt a squeezing feeling in the chest and an aching sensation in both hands, especially pronounced at the end of the fingers, a pressure in the eyes, and heaviness in the head. In another dream that night he saw:

"The raging sea, with a huge black colored wave that's about to overwhelm him."

In fear he woke up concerned about the hard outside wind. In these dreams internal and external stimuli clearly manifested.

In dreams of persons suffering from hysteria, the visual aspect of dreams is more colorfully expressed. Where the dreams of people suffering from neurasthenia exaggerated things, dreams of people suffering from hysteria experienced more reality-like dreams. Additionally, there was a more salient difference: emotional distress, anxiety, fear, tears, and frequent sex scenes. But it can be explained by the fact that in our sample the people suffering from hysteria were mostly young women with a failed family life and an unfulfilled sexual desire.

People suffering from hysteria experience in dreams less thoughts and speech then neurotics in our sample do. But the experience of sound, heat, cold, touch, pain and suffering in their dreams where significantly more common.

Patient C., 24 years old, suffering from hysteria and anesthesia after injury in the cerebellum, has a dream in the night 23/I 1965:

" ..a young man was in someone else's apartment and suddenly heard someone knocking at the door, The man got scared. He went to open the door, the dreaded man entered and hit the dreamer on the head with an axe. The axe broke... The dreamer ran after the man, supporting his head with his bleeding hands ».

In fear the dreamer woke up — with a very bad headache in the place in which he was shot and a nagging feeling in the feet and hands.

Patient I., aged 49, suffering from arachnoiditis[22] hysterical change personality, had a very disturbing dream with an abundance of colors, and often heard ringing or unpleasant sounds. For example, on the night of 4/V 1965, in a dream:

"... in the bath, I see a lot of people washing, all natural colors, I hear the unpleasant sound of scraping against metal basins."

Upon awakening the dreamer was excited and experienced an unpleasant sensation throughout the body, sweating and heat, and was diagnosed with tinnitus[23].

From 1938 on we began to collect dreams from patients with mental disorders at various times. The patients where treated, at the Leningrad Medical and Vologda Regional Psychiatric Clinic. At Kirov hospital we were able to collect dreams from 57 patients. It should be noted that for many patients it was too difficult to obtain information about sleep and dreams. Therefore we worked with this selected sample, keeping in mind that reliability would be doubtful. As you know, one can only get information about dreams after awakening. It is possible that even quite healthy people make some changes in them according to their rules of thinking in the waking state. When awake a mentally ill patient has much more reason to change dreams, because his normal brain activity is broken, so some patients had dreams that should be accepted with some doubt.

---

[22] Arachnoiditis is a pain disorder caused by the inflammation of the arachnoid, one of the membranes that surrounds and protects the nerves of the spinal cord.

[23] Tinutus is the perception of sound within the human ear (*ringing of the ears*) when no actual sound is present.

Based on this, we tried to collect dreams from patients who had a conscious mind and a clear working memory.

Distribution of patients by the nature of the disease and number of recorded dreams is shown in Table 19. As seen in Table 19, mentally ill patients frequently remember dreams, in almost 83% of sleep; in certain diseases they were constantly observed, in others more rare. The hardest thing was to record the dreams from those suffering schizophrenia; these patients did not talk about dreams as much as about their delusional concepts, so it was hard to distinguish the dream from the normal pathological statements of the patient. For example, a well educated patient, aged 32, treated at clinic Vologda, schizophrenic more than seven years (since 1948), expressed delusions of poisoning and heard voices that claimed he was dreaming. However, he reluctantly said that for him it is difficult to distinguish between dream and waking life.

His brother, 30 years old, treated in the same hospital, also with higher education and schizophrenic for more than six years, is often angry and aggressive, complains that: he is burdened by feelings, totally furious and he hears voices, but does not speak about the content. He said that dreams often seem very bright, with the colors related to the whole picture, like in a movie. He is reluctant to talk about the content, saying that his dreams are "just like in life," but he supposedly can not hear voices. The staff proposed he should keep a dream diary. In two months he has recorded three dreams; the first dream in the night from 28 to 29/III 1955: "*Water, granite, beach*"; a second dream in the period between 21/III en I/IV: "*Fell into ...*" and the third dream on the night from 6-7/IV 1966: "*Dreamed of-war*".

This "short" presentation naturally does not allow for any conclusions, but when you consider the fact that stories by such patients lack reliability it becomes clear how difficult it is to learn more about the dreams of schizophrenic patients.

Table 19: Distribution of mental patients who experienced dreaming

Diseases	#. by case	# of sleep	dreams were clear	clear but forgotten	undefined response	no dream
Schizophrenia	4	11	4	-	7	-
Delirium tremens	4	13	13	-	-	-
Chronic alcoholism	7	35	28	3	4	-
Epilepsy	11	60	49	7	4	-
Infectious psychosis	4	12	12	-	-	-
Intoxication psychosis	6	16	14	2	-	-
Reactive psychosis	3	40	37	2	-	1
Encephalitis	4	22	17	-	5	-
Encephalopathy	4	9	9	-	-	-
Psychopathy	2	5	5	-	-	-
Paranoia	2	5	5	-	-	-
Anxious-depressive syndrome of unknown etiology	1	2	2	-	-	-
Vascular sclerosis brain with mental disorders	1	8	4	3	1	-

| Total | 53 | 238 | 199 | 17 | 21 | 1 |

However, other dreams from patients with mental illnesses have a specific character.

Patients with delirium tremens have especially vivid dreams a few days prior to the development of the disease, this disrupts sleep and there are many nightmares: prosecution, scary animals, devils, evil spirits, robbers, frightening scenes, corpses, coffins, war, fire, bugs, mice, rats and other unpleasant pictures. The visual hallucinations in the nightmares were similar for these patients.

Patient C. (from Hospital Gannushkina PB) born in 1918, said after his recovery on 13/VI 1953 that already nine days before the appearance of visual hallucinations he had a terrible dream in which he was chased: *"By a huge animal like a tiger but black, with a big head and horns, and shiny red eyes"* and then these animals and devils began to pursue him in his hallucinations.

A patient suffering from chronic cholecystitis and hepatitis, said on 15/VI 1953 that a few days before the disease he had experienced strange dreams in which:

> *"... he was attacked by beasts like wolves and bears, and it becomes a battle."*

Such dreams appeared during delirium tremens and went on even after recovery. In waking reality his physical and mental state was moderate. His dreams showed ugly scenes of the altered heads of people, animals, injury, damage, large shiny red or glowing eyes, faces that were modified, scary faces of animals, changing to repulsive body parts, and sometimes injury to other body parts.

Particularly striking similar visual scenes were in the dreams of persons suffering infectious toxic psychosis, often appearing before the development of overt pattern of mental disorders. This phenomenon is probably due to the fact that during the infectious toxic disease the whole body suffers, thus there is a significant loss of various organs and systems that very early begin to emerge in dreams.

Knowing the laws of the reaction from somatic stimuli in dreams and sensations associated to them, as reported above, it is easy to understand the appearance of such dreadful dreams about diseases and the visual images associated with the affected body parts and discomfort in them. It is interesting to note that in this disease unpleasant visual hallucinations occurred with nightmarish scenes associated with the same body parts. Such phenomena also occurred in other mental illnesses.

All patients we observed specifically indicated that their dreams were of a bright visual character, but the hearing and feeling quality of the dreams were ambiguous, even if the patient in waking life suffered from auditory hallucinations. During sleep, auditory hallucinations are decreased in almost all patients, whereas visual hallucinations intensified in the evening, especially taking a frightening form in dreams, to the degree that it disturbed sleep. On the exit from a psychosis dreams usually had a calm nature, but nightmares and various uncomfortable situations in the dream still held on for a long time. For example, a mental patient, after the disappearance of his visual hallucinations was in a psychiatric hospital for another two weeks, and all the while his dreaming had an unpleasant character, however, their background became calmer and calmer. It is interesting that dreams of people who consumed alcohol for a long time, often contained visual scenes, involving alcohol, searching for it, attempts to drink it. However, these attempts are usually not completed in the dreams, there is always something going wrong: the bottle is already empty, the wine is spilled, etc. This is especially true in dreams of chronic alcoholics. Speaking of visual scenes involving alcohol in dreams, it should be noted that they are kept in dreams long after apparent healing, and seemingly can serve as an indicator of how much more a patient is craving for alcohol.

We can assume that if such scenes in dreams diminish, the patient is really getting rid of this vicious Instinctual habit. The duration of similar dreams probably depends on the duration of alcoholism and the degree of habituation.

A young man, patient of a psychiatric clinic (1954), used alcohol for about seven years, and claimed that more then two weeks after recovery from delirium tremens in his dreams there were often scenes associated with alcohol, and those scenes appeared less and less.

A pensioner, age 48, alcoholic because of parental disfunction, spent a considerable amount of time, more than 14 years, suffering from chronic alcoholism and was admitted to the hospital for the psychiatric disease for about ten years. A day after the visual hallucinations disappeared he had a dream:

> "*Celebrating New Year with the family, on the table were a lot of bottles and snacks. He was trying to drink, but his daughter took the bottle, saying: 'we will not give you a drink'. He was very upset.*

This dream suggested that this patient did not got rid of his vicious desire, and indeed after discharge from the hospital, he was drinking again. Dreams from persons suffering from chronic alcoholism, usually have an unpleasant nightmarish character, with a sense of anxiety and fear. The patient, we just spoke about often saw dreams in which: "*he fell from a great height, and awoke in fear feeling a slight dizziness*".

Another patient, P., 46 years (1954) and almost IR. (Technical University degree), was a chronic alcoholic for many years. His headaches, often experienced in nightmares involving his head, for example on the night of 16/XII 1954 he had a dream in which:

> "*He participated in the war, he saw shots of fire, many people killed and his head smashed.*"

And on waking he felt severe pain caused by a headache. If on the basis of chronic alcoholism there is a strong delusional concept, it is usually manifested in dreams.

Patient M., 43 years old, from the clinic **VMOLA** (1955), for three years suffering from alcoholic psychosis, expressed his delusions of persecution by claiming that he was allegedly trying to kill parts of his scattered body in the most brutal way to dismember it. On this occasion, he heard voices. He states that in dreams he sees terrible scenes of murder and wakes up in fear.

Another patient in the same hospital, L., 51 years old, suffered psychosis on the basis of chronic alcoholism and cerebral sclerosis* for about a year, not being able to maintain a coherent conversation, said that he has confusing but negative dreams:

> *"Rob him", "carry costumes", "money", "cigarette", and his wife "changes with all the other tenants and sick."*

Of course, we can not rely on the accuracy of reported dreams of only two patients and draw general conclusions, but since almost all similar patients tell their dreams in terms of their delusional concepts, it can be assumed there is some truth in them. It is interesting to note that persons with chronic alcoholism and mental disorders, sometimes indicated sleep disturbance and changes of the content of dreams long before they are diagnosed with psychosis. By in vivo recording seven patients a few months (can not remember exactly how long), five definitely noticed that sleeping and dreaming changed in the course of the disease. The visual dreams, scenes and thoughts were similar to those which manifested in hallucinations and delirium.

The dreams of intoxication dreamt by psychosis patients in the sample are nightmarish, with vivid visual scenes with usually a bad content, characterized by anxiety and fear.

If patients had visual hallucinations, they usually manifest in dreams and, moreover, some

patients pointed out that hallucinations started in a dream before they appeared in reality.

Confusing auditive hallucinations and dreams is rare. We collected the dreams of people without and people with and mental disorders caused by poisoning of quinacrine[24] and tetraethyl lead.

The first case is patient R., 38 years old in 1953, in the psycho-neurological Hospital N2 10 Moscow. Talking about residual effects of intoxication psychosis (quinacrine), said that sometimes she:

"Sees spots on the wall that transform into silhouettes of men and she even hears voices hailing her name."

The patient suffers from severe headaches. And she has disturbing nightmarish dreams. For example, on the night of 17/VI 1953 she had a dream:

"... scary haunted men (similar to those which she saw in the afternoon) hit her on the head with sticks, she woke up in fear."

She had a very bad headache and did not hear voices in this dream.

Patient P. in psychiatric hospital VMOLA (1955), diagnosed with an asthenic condition[25] caused by poisoning in a thermal power plant, had very bright nightmarish dreams. He was not diagnosed with psychosis but the dream is abruptly broken, with lots of unpleasant content. For example on the night of 1956 6/VII:

"... he and his comrades in the service were somewhere exercising. They were attacked by hooligans who started a

---

[24] Mepacrine (INN; also called **quinacine** in the United States and Atabrine (trade name) is a drug with several medical applications. It is related to mefloquine. Its main effects are as an antiprotozoal, antirheumatic and an intrapleural sclerosing agent.

[25] Tetraethyllead (common name tetraethyl lead), abbreviated TEL, is an organolead compound with the formula $(CH_3CH_2)_4Pb$. It was mixed with gasoline (petrol) beginning in the 1920s as an inexpensive octane booster that allowed engine compression to be raised substantially, which in turn increased vehicle performance and fuel economy.

fight, he tried to defend himself, was beaten on the head, and became afraid that they would kill or suffocate him."

In this case we see a reflection of stimuli that occurred during sleep, resulting from the intoxication due to the work in thermal power plants.

Infectious toxic psychosis dreams almost constantly are due to a particularly psychosis. Patient N., 32 years, (VMOLA, 1955), suffering psychosis on the basis of lung tuberculosis, expressed delusions that his wife allegedly wants to poison him, neighbors are watching him, he hears voices inside his head accusing him of crimes. In dreams, he saw scenes associated with this nonsense, but the dreams had no auditive hallucinations.

A patient, 20 years old (VMOLA, 1955), who suffers from a mental disorder after severe dysentery, expressed delusions that he *"fears to be killed"*. In dreams he often sees a black man who speaks to him: "*I'll kill you.*" This patient told me that he first saw this black man in dreams and then while awake.

S., a patient with with vulgar urethritis and psychosis after a rheumatic infection expresses the idea that he is ill with syphilis. In his dreams are often sex scenes of an unpleasant character, these scenes appeared a few days before a mental disorder (VMOLA, 1955). The patient was discharged from hospital after recovery, but on a number of times there was still a health anxiety.

And this is reflected in the following dream: In the night of 19/II 1955 in his dream:

"discharged from the clinic he felt good and healthy, arrived at home, but there comrades kept him away from the troubles."

Perhaps he did not quite recover and other people have noticed and made remarks about it.

For some patients suffering from infectious toxic psychosis it is difficult to distinguish a dream from a visual hallucinations seen in delirium.

They are very similar in content and form, and drift from a night's sleep into the waking state, and vice versa.

Accountant R. was evacuated in 1943 from Siberia to Leningrad with nutritional depletion. There he soon became ill with typhoid fever. The patient remembers the beginning of the disease, then his memory fades on December 10th, when he was seriously ill with typhoid. The subsequent period he remembers quite well with delirium, and high temperature. At this time, he slept 18-20 hours a day and had a lot of dreams. In them he was not an accountant but a chief of food supply from a military unit, engaged in harvesting and distributing food products. When he woke up, the delusions and hallucinations continued in the same vein. He saw a lot of food, ordering their packaging, mailing transport to the various departments etc. This went on for 40 days. On day 41 he woke up in a different condition was and felt "*good, easy as awakening from a deep restful sleep*" without visions, and he felt "*himself*", i.e, an accountant located in a hospital. From that day he recovered mentally. In this case the interesting thing is that the content of dreams, delusions and visual hallucinations was well connected with the life situation and with long-term (chronic) malnutrition.

Mental disorders that occurred with encephalitis[26], manifested in dreams as visual scenes related to the waking state. Patient T., 21 years old (VMOLA, 1954), became ill three years ago: influenza encephalitis, and suffered mental disorders later on like visual hallucinations: he saw a "snake in the belly, a snake with a red and green head" the eyes often seemed "glowing dots" and he heard "voices in his head" which told him to perform certain, often aggressive actions. In dreams, he saw the same terrible scene, "the red or green headed snake's stomach with damage to parts of the body". It should be noted, however, that this patient was high-

---

[26] From Ancient Greek ἐγκέφαλος, enképhalos "brain, composed of ἐν, *en*, "in" and κεφαλή, kephalé, "head", and the medical suffix -*i*tis "inflammation an acute inflammation of the brain.

ly suggestible in questioning, he answered almost always yes, so it is difficult, of course, to rely on the correctness of his stories

In another case, the patient P., 35 years old, from psychiatric hospital Gannushkina PB (1953) after having influenza encephalitis suffered severe headaches for a long time, severe fatigue, drowsiness, he saw nonexistent obstacles: pits, ditches. Sometimes the patient heard voices: calling her to work, knocks on the door or calls. Her dreams were always disturbing, nightmarish dreams in which almost constantly somebody's head was involved - the patients' or other people's head - in unpleasant situations such as: injury, assault, uncomfortable headgear.

Mental disorders, manifest in dreams by way of visual scenes that express the content of the mental disorder and the situation that caused them. The same type of recurring nightmares and thoughts in dreams, if not associated with a somatic disease that it is easy to eliminate, should always alert the psychiatrist. This provision is of particular value with slowly developing psychiatric disorders, weakly expressed on the fringes of normal and pathologic condition.

For example, an elderly man V. said (5/X 1956) that during the day he sometimes get scared of being alone in the room. This man often dreams this fear:

> "He found himself alone in the darkened room, the walls of which begins to move, converging to such an extent that he was nearly crushed, and he wakes in fear."

Thus, even this brief examination of dreams of psychiatric patients gives an indication that dream content is definitely associated with their associated clinical psychopathology and reflects the peculiar characteristics of the mental disorders. This position is noted by many authors (Krauss, 1858; V. Griesinger 1881; Yu Janet 1911; V.H. Kandinsky and y, 1890; A. Kronfel Ld, 1940; V. A. Giljarovsky 1951; R. Dolmirsky 1961 etc.) and used for analysis of psychopathological manifestations of a number of mental illness.

Of special interest for the understanding of the working of the brain during sleep were the dreams of persons suffering from brain tumors and other local organic diseases. We observed similar patients in the mental Institute Bekhterev (1941, 1964, 1965 etc.), Lviv Regional Psychiatric hospital (1952.) In the 1st Faculty Surgical Clinic (1954) and the neurosurgical clinic (1959.) Kirov Vmola, as well as the Russian Scientific Research Neurosurgical Institute. Prof. Polenova (1963 - 1966) etc.

The nature of brain injury and the frequency of occurrence of dreams are shown in Table 20. Dream recall from patients suffering organic brain lesions is much lower than in patients with neuroses. If you compare the 238 sleep observations recorded only in patients with tumors, dreams were clearly remembered in 109, i.e., about 45.7 %; were clear but forgotten in 29 cases about 12.1%; a vague answer was obtained in 65 - about 27.3 % of the cases, and in 35 cases, i.e., approximately 14.7% patients said that they did not see dreams.

Tabel 20
Organic brain disease, and the frequency of dream appearance

Disease	# of cases	# of cases explained	# dream remembrance			
			is clear	clear but forgotten	indefinite response	were not
Swelling of the occipital region	1	12	3	2	7	-
Tumor of the left parietal-occipital region	1	2	2	-	-	-
Tuberculosis left occipital region	1	3	-	1	1	1
Tumor at the junction of the temporoparietal and occipital lobes right	1	6	1	1	4	-
Tumor front temporal left parietal region	3	17	4	-	4	9
Swelling of the deep divisions in the right frontotemporal region	2	5	3	-	2	-
Tumor of the left fronto-temporal region	1	13	5	2	3	3
Pituitary tumor	4	30	7	4	7	12
Hydrocephalus	3	18	18	-	-	-

Tumor of the left parietal area	2	11	-		4	5	2
Tumor of the left frontal region	3	10	4	-		6	-
Tumor of the right frontal region	2	12	5	4	3	-
Tumor of the right parietal area	1	9	7	2	-	-
Parasagittal tumor on the right hemisphere	1	3	-	-	3	-
Tumor fronto-temporo-parietal region of the right	1	6	1	-	5	-
Swelling of the deep divisions midline with spread into the lateral ventricles	1	10	7	-	3	-
Suprasellar tumor	1	10	1	-	5	4
Swelling of the 4th ventricle	1	3	1	-	1	1
Swelling of the third ventricle and basal ganglia	1	2	1	1	-	-
Base of the brain tumor	5	31	25	-	3	3
Swelling of the foramen Magendie	1	3	-	3	-	-
Swelling of the midbrain	1	2	-	-	2	-
Neuroma of the right auditory nerve	2	5	4	1	-	-
Left acoustic neuroma nerve	1	2	2	-	-	-

Trigeminal neuralgia branch 2 and 3	1	1	-	-	1	-
Tumor of the cerebellum	2	24	18	3	1	2
Injury	1	24	19	5	-	-
Through a bullet wound of skull with complete destruction both optic nerves and damage to both frontal lobes	1	6	5	1	-	-
Cerebral aneurysm	3	10	10	-	-	
subdural saccular hematoma and frontoparietal the left temporal region.	2	6	4	-	-	2
Stroke in the right hemisphere	1	30	28	1	1	-
Stroke in the left hemisphere	2	13	8	1	2	2
Arachnoiditis brain different localization	16	117	101	11	1	4
Condition after Removal foreign body in brain	1	2	2	-	-	-
Brain commotion encephalitis	5	35	29	2	3	1
Encephalitis and meningoencephalitis	8	37	32	-	5	-
Total	**86**	**533**	**360**	**49**	**78**	**46**

Many patients indicate that they often saw the whole picture of progress of a disease in dreams. Dreaming increased particularly at the beginning of the disease, and the content of the dreams changed with time. As the patient gets better these kind of dreams become less frequent. It is interesting to note one fact: very often when there is a significant increase in intracranial pressure in a patient, remembering a dream becomes less common. This phenomenon is probably due to a sharp decline in brain activity due to weakening of the underlying neural processes of excitation and inhibition. Increased intracranial pressure syndrome occurs in marginal consciousness. Furthermore, as mentioned above, a decrease in dreams was noted in all patients suffering from a disease in which there was a general lack of of bodily strength; i.e, in the case when there is a weakening of the basic nerve processes (asthenia). In this connection it is interesting to note that in the experimental study of imaginative thinking (having a certain attitude and dreams) it was observed that the significant feature of asthenia in humans is that figurative representations are absent (L. G. First, 1964). This could explain this phenomenon as a weakening of nerve processes. F. Slichevski (1962), suggests that a decrease in the number of dreams of persons with hard organic brain damage was due to the fact that during these diseases there are no marked hypnotic phases. Considering dreams as analogous to hallucinations and their hypnotic phases, he believes that due to weakening of the basic nervous processes of excitation and inhibition that hypnotic phase-states do not occur in persons with serious brain injury and therefore there is no dreaming. As seen from the foregoing, the weakening processes of excitation and inhibition (or, in other words, decrease in the activity of the brain) determines the appearance of hypnotic phases and dreams. But the role of dreaming during disease of the brain needs to be studied further because even though Slichevski did not observe these hypnotic phases, there are still dream-like visual hallucinations and delirium in patients that suffer from extensive damage of the brain.

Of particular interest is the question of how dreams of patients with lesion in various areas of the brain change. Here it was important to observe the impact of the failure of a functional area on the appearance and nature of dreams. The majority of patients we observed had a tumor removed, so the localization, lesion size and nature of the tumor was established quite accurately, allowing more definite associations of certain types of dreams with structural brain formations. However, even with significant loss of various functional areas a clearly expressed change in dreams is not marked.

We were not able to observe patients with extensive bilateral loss of visual parts of the brain causing total blindness. That probably would have been of value for judging the appearance of visual elements in dreams.

We observed seven cases of lesion of the occipital area and the visual pathways and neither of them gave evidence to draw definitive conclusions. In all seven patients there was cortical vision impairment (Cortical visual impairment (CVI) is a form of visual impairment that is caused by a brain problem rather than an eye problem). It is necessary to note that complete blindness, the loss of the function of the whole optic cortex is an extremely rare phenomenon.

According to E.P. Kononov (1926), and N. S. Preobrazhenskoy (1950), E. J. Troano (1955) and others, the destruction of the calcarine fissure[27] in both occipital lobes usually diminished the central macular vision range of 5 - 100. More destruction of the center yellow spot of the retina located at the poles of the occipital region leads to total blindness. Bilateral cases of persistent blindness in lesions of the occipital lobes are rare.

During the Great Patriotic War a case of complete blindness was noted by N. S. Preobrazhenskoy (1950), only one time in 60 patients with injuries in the occipital and parietal-occipital region. It

---

[27] The calcarine fissure is located on the medial surface of the occipital lobe and divides the visual cortex into two.

is also necessary to remember that each functional area, apart from its center, is across a large area of the brain, a phenomenon also noted by I.P. Pavlov. The latest electrophysiological and anatomical studies of the brain, as well as clinical observation confirm the position taken by I. P. Pavlov on the existence of extensive neurological network of the visual area in the brain (Sariskov, 1949, 1964, G.V. Gershuny and A.V. Fine, 1949; L.A. Kukuyev, 1955 A.M. Greenstein, 1956; Livanov, 1962 I. N. Filimonov, 1964 L.B. Litvak, 1964 M.S. Lebedinskii, 1964; Kao-Liang, 1950; Penfild, 1954, 1958; Magoun, 1952, Jasper, 1949, etc.). Although some foreign authors came to unfounded statements of the predominant role in the activity of the subcortical brain formation, some of the evidence they presented can be seen as new and interesting facts for understanding the nerve system. The lack of strongly pronounced dreams in patients with lesions in various functional zones are probably due to the fact that each zone has a significant range in the human brain. In several of our cases, this phenomenon is due to the fact that the functional zone was destroyed only partially or in one hemisphere of the brain. The question of localization of the embedded brain is very complex and far from being resolved. Nevertheless, results of experimental studies and clinical observations of recent cortical areas for different function detects the same function. Let us consider the evidence relating to dreams from patients with tumors of the brain.

The noticeable deviation of dreams of healthy people from patients is that the patients have less positive dreams. In them, they do not celebrate. The dreams are filled with visual images and scenes, other elements were rare. Their dreams resemble reality. On the night of 29/I 1960 a patient dreams:

"... I am in a bar with drink s like vodka and beer. Then a fight starts, I try to get away, but can not find the door, I hear a voice declaring that I have to find a way out.

People around me walked and ran to the right side and disappeared all at once. Then I hit someone on the head with a bottle."

Upon awakening the dreamer had a headache and a dry mouth. In this case, the dream showed the headache, pain, irritation in the mouth and impaired vision.

Another patient, G., engineer, 60 years old, was admitted to Neurosurgical Institute Polenov on 18/I 1962 with complaints of headache, vomiting, and blurred vision. In August 1962 this patients was diagnosed with a tumor at the junction of the temporal-parietal and occipital lobes. On 4/XII 1962, an osteoplastic trepanation was made in the right temporal parietal and occipital region, bone flap removal and biopsy of the tumor. Microscopic examination of the tissues showed that the tumor probably metastasized[28].

We observed the patient five months after the onset of illness. The patient was lethargic, sluggish, reluctant to answer questions, but basically correctly oriented in time and place. Sometimes the patient had visual hallucinations with a bizarre character like *"scary faces of animals", "bloody heads of people", "devils"*, etc.

The patient had a lot of scary dreams at first (war, fight) and then less, but did not remember the details. Currently sometimes the patients has scary dreams, for example, on the night of 13/I 1963: *"Got in from where he was beaten on the head, then thrown from the mountain into the abyss"*. In fear the patient awoke with a very bad headache. The patient says that the dreams sometimes displayed the same "terrible visions" he experienced in December, while hallucinating. If we consider these stories reliable, we must admit the idea of similarity between dreams and visual hallucinations.

Because the patient lost vision due to atrophy of the optic nerves, this case indicates that dreams and hallucinations can occur wit-

---

[28] Once the tumor is formed, cells may begin to break off from this tumor and travel to other parts of the body. This process is metastasis

hout the participation of the eye. The literature describes the spectators hallucinations in blindness: Charles Bonnet Syndrome[29]. Flinn (1962) cites the case of a 75-year-old blind woman with visual hallucinations who is observed for 8 years.

Patient D., technician 53 years old, with left occipital lobe tuberculoma entered Neurosurgical Institute Polenov on 2/1 1963. D. got ill in1962, there was a headache, dizziness when walking, and a decrease in visual acuity. D went to visit an optometrist that specialized in complicated patients. The Optometrist diagnosed a central neuron lesion of the optic nerve in the left hemisphere. The patient's condition is lethargic, cooperating reluctantly, speaks slowly, individual fragments of phrases, but responds to questions correctly. D. rarely remembers dreams, only that the content is *"terrible but, without stories"*. During the week we were unable to record any dreams. In this case there was a pronounced reduction in brain activity during wakefulness and, of course, during sleep. That is the most probable reason that dreams were rare and not remembered.

The next patient whose dreams we will discuss has optic nerve damage on both frontal lobes. A., 29 years old, was admitted to neurosurgery clinic VMOLA on 20 /I 1959 concerning a diametrical injury on the skull. We observed the patient 50 days after the injury. Most of the time he was sick in bed, asleep or dozing. A. was time and space oriented. Sometimes the patient expressed delirium: "his wounded enemies." Visual activity in both eyes is zero. We agree with the optometrist about the complete destruction of the visual nerves. He had a lot of dreams, most of them were unpleasant, often the patient sees scenes of his Injury, ie, the situation in terms of his delirium. For example, in the night of 12 /

---

[29] There's a relatively obscure medical condition called Charles Bonnet Syndrome. It happens to people who are either losing their sight, or completely blind. Suddenly, one day, they see vivid and detailed hallucinations. It's been documented since the 1700s, and even today nobody is quite sure why it happens

XII in 1959 in a dream "... *had to deal with a specially problem. He was attacked by bandits and wounded on the forehead.*" That night he had a headache.

In his dreams visual scenes are often gray, but sometimes he observed other colors like - yellow, brown and green. Because of his injury the patient was not able to tell something about his dreams. Sometimes there were dreams, but the content was not remembered. This event, like others described above, show that the destruction of both optic nerves do not change or the character of dreams, ie confirms cortical origin of visual elements and even the colors of dreams. This is also confirmed by the following example.

Patient P., 48 years old (Bekhterev psychoneurological research institute, 1965), has completely lost sight in his right eye four months and two weeks ago ago on the left eye. At the time of the examination he lost vision in both eyes. This patient complained of headache, pain, drowsiness, dreaming, and in his dreams seeing as before, but only „terrible visions" (of dead, the war). Because the medical condition of the eyes was not the result of the occurrence of blindness, the question of determining the localization of lesion of the visual pathway became relevant. And the presence of visual dreams express the idea that blindness in this case is not associated with a decay of the occipital lobe, as in complete blindness, there could be visual images in dreams that originate from the cortex. Indeed, a tumor in the right temporal region was found during operation, a compression of the chiasm opticum[30]. The tumor is not removed. After surgery, vision is not improved, but the patient continues to see scenes in dreams. For example, on the night of 21 / V 1965, (on the eighth day after surgery) in a dream

"... bought products (bread, cereals.) add up in a bag, but remembered that he sees, and burst into tears."

---

[30] The part of the brain where the optic nerves (CN II) partially cross.

Upon awakening he feels a little more heaviness in head and eyes.

As noted above, with the deterioration of the occipital lobe noticeably more changes occur in the visual background of dreams.

Patient M., 38 years old (Bekhterev psychoneurological research institute, 1965), noted brief temporary loss of vision, this usually occurs after severe headache attack, pain mainly in the neck, sometimes diplopia. Prior to this, the patient was examined in Leningrad, with a diagnosis of multiple sclerosis and arachnoiditis in the posterior cranial fossa[31]. Vision in both eyes was 1.0, ocular fundus normal, bipolar visual field. The patient had a disturbed sleep pattern. In his dreaming the patient has vision, but sometimes "loses" sight during sleep, as well as in reality. For example, on the night of I8 / IV 1965 in a dream:

> "... came to the reception to the clinic to visit a neurologist, who had treated him before. Got into an argument with the doctor, and said to him: "What have you done! If you had treated me earlier this would have been temporary, now am blind. "The doctor promised to provide a consultation from other experts."

At awakening the patient has severe pain in the neck, and he was unable to perceive daylight, although it was already morning. This dream manifested not only pain in the neck, but probably the condition is temporary and the loss view has a cortical origin. It is interesting that at the onset of temporary blindness the patient has not stopped dreaming, he thought about the disease, talked to a doctor, but his dreams did no longer have clear visual images. Confirmation of the direct communication of visual images exactly with the visual field is the also by the fact that the cortical he-

---

[31] The posterior cranial fossa is the rearmost hollow or depressed area in the base of the cranium, which constitutes the upper part of the human skull. It is also the largest and deepest area of the skull. The posterior cranial fossa is one of the three cranial fossae, the others being the anterior cranial fossa and the middle cranial fossa. Like the other depressed areas, it bears the lobes of the brain.

mianopsia, (blindness (anopsia) in half the visual field of one or both eyes, usually on one side of the vertical midline) in dreams the same fields of view are not seen.

The above observations provide some data to analyze the visions in dreams dreams as a guide to damage in the cortical brain structure. In particular blindness, evolved as a result of intracranial disease (in the absence of pathological changes in the eyes). The disappearance of visual images in dreams points to the total destruction of the visual cortex, the presence of the same visual dreams in this case indicates the preservation of the visual cortex. When there is a tumor in other areas then the visual cortex there are some changes in the nature of, or in the content of dreams.

It is noted for example that if a tumor is located in the left parietal region and if it is of a considerable size, the patients dreams rarely had signs of these tumors. Such patients usually told us they had dreams but forgot the content. We tried to record the dreams of five such patients,over 27 nights and they reported only four dreams.

Here are some of the data in these patients.

Patient K., 21, admitted to the neurosurgery department of the faculty of the Surgical Clinic VMOLA in Kirov on 12 / II 1954 with a diagnosis of a tumor in the left parietal region. K. got ill three and a half months ago. In the beginning there was a headache, then convulsive seizures, central facial palsy, right sided hemiparesis, impaired speech by type of motor aphasia. When examined we found sided hemianopia. This patient was operated on 5/III, during which a tumor size 6x7x2,5 cm in the upper layers of the left parietal area of the brain was removed. The postoperative period was quite satisfactory. We observed the patient 17 days after the operation. At this time he was sluggish, spoke slowly, was inarticulate, and maintained his right hemiparesis. From the words of the patient we could deduce that early in the disease he had disturbing dreams with unpleasant content, but now he does not remember dreaming. He sleeps poorly with less dreaming or memories of dreams. After the surgery sleeping became better but the patient does not remember dreams.

The second patient, M., 21, a mechanic of a locomotive depot, re-entered into that same clinic on 29/1 1954, with a diagnosis of a recurrent tumor of the left parietal area. The first time he was at the clinic on 28/IX 1953; at this time he was operated to remove the tumor. The re-operation was done on 12/II 1954. In the center of the parietal region was a partially degenerated tumor contai-ning a yellowish colored liquid. The tumor occupied the center of the parietal lobe and went deep into the brain to the basal ganglia and the junction of the temporal and occipital lobes. The surgeon is under the impression that the tumor is completely removed. Tumor size 6x6x4 cm, weight 98 gram (3.46 ounces), the posto-perative period was satisfactorily. We observed this patient forty days after the operation. At this time he had a paresis of the right hand, and periodically complained about upcoming headaches. On the part of intelligence and speech there are no big changes noted. The patient said that at the beginning of the disease he was disturbed by alarming dreams, the had a lot of dreams with a frightening character, but now he can not remember them. Sleep remained sensitive after the first and second operation, but after the first operation he remembers dreams and after the second, he does not see or does not remember dreams. Other lesions of the left parietal and adjacent areas, such as trauma and hemorrhage[32] reduced bad dreams and the memory of them as well.

Lesions in the right parietal region have a different affect on dreaming. In such cases, patients would definitely indicate that they dream a lot and remember them well. Recently, in connecti-on with studies of mental disorders in local brain lesions it was observed that a wound in the right hemisphere of right-handers more frequently leads to chronic psychoses than an injury in the left hemisphere (M. Lebedinskii, 1964).

---

[32] Bleeding in the brain.

Perhaps these data and our observations are indicative of the same phenomena. Anyway our observations confirm the view of M.S. Lebedinsk (1964) and G.P. Gubkin (1964) that the right hemisphere is not as "silent", and has some relevance to mental activity[33].

In one case, we observed a dream patient K, 25 years old (VMO-LA, 1954), with skull and bone tumors in the right parietal area, growing outwards and inwards size 8x8x6 cm; the substance of the brain tumor squeezed the right parietal region, without damage to the dura mater. The patient complained of a headache, more in the right half, had unpleasant sensations in the left arm and leg. Dreams before and after surgery were alarming, with an abundance of unpleasant content in which he damaged his head on the right side; it was especially noticeable in the first days after surgery. We recorded his dreams before and after surgery. Nine cases in seven nights sleep were vivid dreams and only in two cases the dreams were forgotten. Thus on 20/III 1954, he learned that he will be operated, and he was thrilled by this message. The patient could not sleep the night before the operation. In the subsequent part of the night he had disturbing dreams, many with nightmarish content: of war, and fighting. In the morning the patient had a headache, and experienced general weakness. Day 21/III the alarming condition continues, but in the night he takes Luminal — 0,1[34] and slept peacefully, and remembers his dreams.

> After surgery the night of 22 to 23/III he has restless sleep, headache, but does not remember dreams. Day 23/III the patient has a headache, especially in the area of the wound. Sleep in the night of 23 - 24/III was more relaxed even though there were plenty of bad dreams: quarrel, fight, war and in all of them he hurt his head; in one dream:

---

[33] Currently there is some evidence found for this, for example this book: *The right cerebral hemisphere and psychiatric disorders*. Cutting, John New York, NY, US: Oxford University Press. (1990).

[34] Luminal is a medication used to control seizures.

"he was wounded at the right half of the head, in another he was beaten on the right side of the head" and every time when he feels pain when he wakes up.

However, in another case, patient P., 50 years old (Institute Bekhterev,1965), the tumor is in the frontal-temporal-parietal region on the right, located at the base of the brain in the anterior and middle cranial fossae. In this case dreams are rare but the general condition of the patient was cumbersome and most of the time she was asleep or dozing. During 10 days of observation she reported only one dream.

With a large hemorrhage in the left hemisphere in the area of the middle cerebral artery, accompanied by hemiplegia and disturbed speech, even many months after stroke and recovery of speech and motor functions, patients said dreams are rare and poorly remembered. One patient with hemorrhage in the same region of the right hemisphere shortly after an acute period of minovanii[35], noted dreams quite often and remembered them well.

We observed a patient A., 41 years (Institute Bekhterev 1941), who had a stroke caused by thrombosis in the right hemisphere at the middle cerebral artery. Thirty-seven days after he his stroke he had a light left-sided hemiparesis. He walked in on his own, a bit unstable. He recorded twenty-eight dreams. Dreams often wore an unpleasant character, but in the nightmarish scenes it was difficult to notice the specific features of the affected areas of the brain. More often than healthy people he dreamed about difficulty of moving his left hand and leg.

When after a stroke in the left hemisphere speech is restored patients begin to notice the occurrence of dreams. Patient G., 31 years, about two years ago suffered a stroke in the left hemisphere, with complete loss of speech and right hemiplegia. Entered in the Institute Bekhterev 24/II 1941, complaining of headache and pain.

---

[35] Period of minovanii = a lethargic state of delirium.

Neurologists observed residual effects right hemiparesis and mild dysarthria[36]. The patient spoke quite clearly and wrote. About his dreams in the first months after stroke I could tell you nothing definite, but then they appeared (exactly how many months without dreaming I do not remember). Dreaming has recently been fairly frequent with visual scenes of *"falling through the ice,"* *"falling into the water,"* and unpleasant scenes and thoughts associated with the affected limb. During the observation period, the patient recorded eight such dreams. However, in this case, the nature of the injury is not captured by the dreams.

With a frontal lobe tumor decreasing imagery in dreams is not observed in the data we have collected. Patient F., 39 years old, entered the hospital VMOLA 11/II 1954 became ill in 1950, first headaches appeared, involuntary urine emission at night. In the beginning of the disease he slept soundly, and does not remember dreaming. At the beginning of 1954, his health deteriorated sharply, headaches increased, disturbing dreams appeared with unpleasant content: falling off a cliff into the abyss, fighting, war, beatings, head injury, a bull with big horns stalking him, etc.

He was operated at the clinic on 8/III 1954, during which a parasagittal tumor was removed. The tumor occupies the entire medial aspect of the right frontal lobe, its size 5 x 5 x 3 cm. We observed this patient twenty days after surgery. Orientation in time and space was not treated, sometimes he expressed delusional thoughts of a sexual nature related to his life. In eight nights he recorded only one dream. In this dream no significant changes noted. In the afternoon of 11/IV1954, his health was good, almost no pain in the head, the patient went for a walk. His brother came to visit him. On the night of 11 to 12/IV he slept well, there was one dream in which:

---

[36] Dysarthria is a motor speech disorder resulting from neurological injury of the motor component of the motor-speech system.

"... he got home, I saw my wife, children, brothers, everyone is happy about his return from the hospital, and he too was glad wanted to drink wine, but feared that he would get sick again".

Yesterday, during a meeting with his brother he talked about his family, wife, children, and also that it would be nice to drink wine.

Patient T., 48 years old (Institute Bekhterev, 1965), had a tumor removed. The dimensions of the tumor 4.7 x 3, 7 x 2.5 cm from the posterior portions of the right frontal lobe.

The tumor was associated with the inner surface of the dura brain membranes, and pushed into tissue to a depth of about 5 cm. After surgery this patients was often sick in his dreams. For example, on the night of 10/III 1965 (23 days after surgery) in a dream:

"... the attending physician gave me a massage on the right half of my head, the place where there was an operation, then he puts this part of the head back in its place, saying that "now Nina Timofeyeva can get up," and this time squeezed my left hand."

Upon awakening the dreamer had a headache in the area of the operation, was in a convulsive state and had the sensation of pain in the left hand.

Patient C., 18, a student of 10th grade, was admitted to neurosurgery clinic VMOLA 21/XI 1959, complaining of a headache in the frontal region and weakness in the right arm and right leg. The patient got ill two years ago. The clinic diagnosed a deep tumor in the left frontal lobe and limited encephalitis[37] frontal localization of the left, with weakly pronounced symptoms of right-sided hemiparesis. The patient was not operated on. The patient was unstable, euphoric and insufficiently aware of his condition. When the

---

[37] Encephalitis is an inflammation of the brain, usually caused by a direct viral infection or a hyper-sensitivity reaction to a virus or foreign protein.

patient had no headaches she had unstable sleep — shallow, with an abundance of dreams. When the patient suffered a headache there was sleep without dreams.

The patient said this about the different content of dreams:

"In the dream, I see everything that was on the eve of the day" or "what they wanted to see" "I want to see mom or dad - I see them in my dreams."

However, in many dreams she observed visual scenes related to the head: *"different freaks with large or, conversely, smaller-headed, scary animals with large heads and horns, sometimes bruising the head, or falling from the mountain."* Such scenes began to appear about 3 years ago and appear usually in the night before a headache attack. Recently the patient began to dream about figures linked with the right hand and right foot. For example, in the night of 17 / XII 1959 in a dream:

"She was attacked by boys, beaten on the head with branches she tried to run, but could not move her right leg, she wanted to defend herself, but the right hand did not move too."

When she woke, she had a headache, and her right hand and right leg was numb.

As the condition of a patient worsens, dreams get more rare. An example, is patient K, 29 years old, hospitalized in the same clinic on 1 / III 1954 with the diagnosis: suprasellar tumor[38]. The patient had an operation were the surgeon discovered a more solid tumor on the basis of the right frontal lobe and right lateral ventricle. The tumor was not removed. The patient's condition quickly deteriorated, but he had a lot of dreams, and if you can rely on his words, his dreams are not visual. Only once he told a dream in which: *"he was at home, in Siberia, saw fields, meadows, fields"*.

---

[38] Above the sella turcica: a depression in the base of the skull where the pituitary gland is situated.

All patients seriously ill with different tumor localizations mention a reduced number of dreams.

Patient F. aged 46 was hospitalized at neurosurgical clinic VMO-LA, 1959, and operated on 26 / X 1959. In the posterior portions of the left frontal lobe a tumor is detected, extending deep into the brain. The tumor measures 8 X 6 cm and is removed completely. We observed the patient two weeks after the operation. The patient seems euphoric and insufficiently aware of his condition. The patient spends most of his time in bed, sleeping or napping. He tells us he does not dream. Before the disease he had dreams but he could not recall them.

Patient C., 38 years (Institute Bekhterev, 1965), very rarely dreams. C. is diagnosed with a tumor deep in the divisions of the right frontal lobe. The patient sleeps most of the day.

Second patient from the same institute, SA, 45 years, after surgery for tumors in the left deep temporoparietal area also stated that he sleeps soundly and rarely dreams. And this patient was most of the time sleeping, and often had seizures.

The patient M., 46 years old (VMOLA, 1959), during surgery meningiomas are removed from the left frontal region measuring 4 x 3 x 5 cm. After surgery the patient suffered from apraxia, general lethargy and a sharp decline in memory function. The visual acuity was low: D= 0.05, S= 0.3. Most of the time the patient was asleep. He could not tell if he had been dreaming.

The same phenomenon with respect to the frequency of occurrence of dreams is noted in patient K, 38 years old, from the same hospital with a tumor in the right hemisphere and parasagittal region. Most of the time the patient was in bed — sleeping or dozing. The patient is hardly conscious. Correctly oriented in time and place. Delirium and hallucinations expressed. His memory dramatically reduced at times and he is euphoric. He has no dreams he can remember.

Patient N., 52 (Polenov Neurosurgical Institute 1963), with a tumor creating a biological interface deep in the divisions of the frontotemporal lobes of the right noted a reduction in the number of dreams. The surgeon made a resection trepanation in the right frontotemporal region and cyst aspiration. The patient's condition was severe, most of the time he was in a drowsy state.

In another case, patient M. with a tumor in the brain base, was in a euphoric state, said that he did not dream (1952). However, four other patients with tumors of the brain base front, middle and posterior fossa, a disease much less severe indicated they often have dreams. In twenty-nine cases of sleep they noted twenty-five dreams. Those dreams were, as a rule, bad in character, the content was associated with the clinical symptoms of the illness, but without apparent changes in their structure.

A decrease in the number of dreams is noted in severe diseases, like patient R., 22 years old (in 1952), with a tumor in the fourth ventricle. Patients with pituitary tumors hardly ever remembered their dreams, or rather these patients often said they have dreams, but they forgot their content. This dependency of the frequency of occurrence of dreams on the severity of disease was also observed in patients with pituitary tumor. Indicative in this respect is a patient, 34 years (Institute Bekhterev, 1964), we observed after a pituitary cyst was removed. The patient was deaf and slept most of the day. Sometimes during the afternoon and night the patient entered an excited state, screaming, *"Help! Robbers! Help!"*. The patient said that sometimes she has terrible dreams, but forgets the content.
There is a loss of vision of half of the field of view and the arteries of the temporal lobe became narrow; swollen. However, after three months of being in this state she significantly improved: became less sleepy, started to work again, and at the time reported several dreams. She said that before the operation the robbers always haunted her, and now only in dreams. For example, on the night of 5/II 1965, in a dream:

"was sitting in a large dark room, dozing, but I saw that the robbers cut nurses in half, then they tried to kill me."
One robber held a large knife to her throat: *"I cried, 'Help! Save!' Then came our helpers and killed many thieves."* The dreamer woke up in fear with an unpleasant feeling in the eyes, a sore throat, headache, and an aching sensation throughout the body.

A neighbor in the ward said that she just screamed "help, save me!".

Another three patients with pituitary tumor, we observed immediately after surgery, their condition was severe: stupor, drowsiness, dreaming is rare: for 28 nights only five times a dream was written down. The content of the dreams is almost the same as in healthy individuals. However the content clearly reflected their symptoms.
However, some brain damage-dreams are frequent and are very bright. For example, with edema of the brain in three patients we observed good dreams every night. General condition of patients was much better than those described above.

As an example, patient K., 15 years (VMOLA 1954) hydrocephalus with partial occlusion[39]. There was surgery performed in the posterior fossa arachnoid to remove adhesions. We have observed the patient nine days after surgery. He complained of a headache, especially in the occipital region, nausea, was often vomiting, his sleep is very anxious, intermittent and always with an abundance of nightmares: a kind of rotten fish, meat, fight, attack on a wolf, big mice that bite the hands, feet, and very often the head. In the night from 25 to 26/IV the patient was sleeping late, sleep was intermittent, there were many dreams with offensive content. In one:

---

[39] Occlusive hydrocephalus: water on the brain.

*"suddenly in the wards wolves appeared and began to at-
tack him, biting his arms, legs and head. Really scared."*

The dreamer even woke up, looked in awe, whether in there was a
wolf in his house, but there were no wolves what was very happy.
The dreamer experienced headache and aching hands.

Similar to this case history of diseases is patient M., 36 years old,
in the same clinic, suffering from swelling in foramen Magendie.
M. got ill ten years ago. At first she noticed a change in sleep, she
had a lot of unpleasant dreams, at the same time M. experienced
headaches. For ten years, she slept only 4-5 hours a night. Her
dreams are often about cemetery, dead, her dead father who drag-
ged her into the grave, freaks with altered heads, clutching head,
arms or legs, she often wears a hat or feels uncomfortable blows
to the head.

Dreaming observed and subcortical lesions nodes in the region of
the third ventricle. In the Institute Bekhterev in 1965 we saw the
patient I., 41, who had Thorkildsen's surgery[40] two years ago (The
tumor is not removed). At the time of observation he took radio-
therapy felt satisfactorily, slept well, was dreaming, but often did
not remember them.

Another patient K, 15 years old, from the same clinic; prognosis
tumor with average diameter, we tried to make a clinical observation. In the period of observation the patients' audible capabilities
diminished, sleeps soundly most of the day, he does not report
dreams.

Patients with lesions of the cerebellum frequently reported
dreams. We observed two patients after tumor removal of astrocy-

---

[40] Thorkildsen's surgery: technique for operative treatment of hydrocephalus
(ventriculocisternostomy, Thorkildsen's shunt). His technique was the first suc-
cessful procedure for shunting of cerebrospinal fluid and soon became interna-
tionally known and accepted as the standard operation for obstruction of the
aqueduct or posterior third ventricle.

tomas[41]. In one case, the right, the other left hemisphere) and one after injury of the cerebellum. The first patient recorded18 dreams in 24 nights, in the second patient 19 dreams in 24 nights. The structure of their dreams differed little from dreams of healthy people, but there were clearly elements in the dream that referred to their medical condition.

A reduction in dreaming is noted after severe injuries to the brain. For example collective farmer T., 45 years old (VMOLA, 1959), a year ago suffered a head injury with loss of consciousness, after which he formed extensive subdural hematoma saccular, occupying the entire left parietal lobe, with partial transition to the temporal and frontal lobes. Before the operation the patient was lethargic, about dreams he could not tell.

After surgery, the patient's condition was improving rapidly, and already in the second week he began to notice dreams. In one, *"he was driven to a stroller bandaging, stroller skating, and he fell with it."* Two months after surgery the patient did not express any agony. Dream were satisfactory, and were of a calm nature. A similar condition was also observed in another patient, J. , 58 years (Institute. Bekhterev, 1964 - 1965), subdural hematoma after injury in the left fronto-parietal region. In the period immediately prior to the deterioration and after surgery he was deaf, he slept much but did not remember dreams. As soon as his condition improved after surgery he began to notice dreams.

In two cases of concussion after the patient returned to consciousness, dreams with positive content were remembered. M., 29 years, we observed after surgery; during which a chip from the blade of a knife that penetrated into the right parietal-occipital region was removed, at a depth of 3 cm it. After surgery sleep was satisfactory with an abundance of dreams. The content of his dreams often displays his head or heads in general in unpleasant situation (beatings, injury).

---

[41] A type of cancer of the brain, They originate in a particular kind of glial cells, star-shaped brain cells in the cerebrum called astrocytes.

We observed two patients with an aneurysm of the left internal carotid artery[42] and one with an aneurysm of the circle of Willis[43]. Dreaming of all these patients was quite often nightmarish (war, dead, disease), but each patient had some features associated with clinical observation of the disease. The structure resembled dreams of healthy people, they are almost indistinguishable.

In three cases, we observed patients after surgery for a tumor in the auditory nerve in the cerebellopontine angle, in two cases on the right and one on the left. These patients often had dreams scenes associated with vertigo, tinnitus and headache, such as motorcycle ride, riding on horseback in the American mountains, falling from an airplane, beatings of the head and so on. In one case, patient P., age 34, diagnosed with the shingles and a disturbance on the trigeminal nerve area complained of pain and burning in the crown part of the left of his head. In a dream P. was shown: the walls of the "*separation of the parietal region, just on the left side.*"

Patients suffering from arachnoiditis[44] dream frequently. In sixteen patients with relatively mild disease we collected 101 dreams for 117 nights. Most of them recorded the dreams themselves. Sleep in all patients was unsettling, dreams were very bright and, as a rule. unpleasant or nightmarish. The content of those dreams quite clearly reflected painful sensations.

Thus, patient P., 40 years old (Institute. Bekhterev, 1965), who suffers from cerebral arachnoiditis, complained of constant head-

---

[42] An aneurysm is a balloon-like bulge or weakening of an artery wall. As an aneurysm enlarges it puts pressure on surrounding structures, causing headache or vision problems, and may eventually rupture. A ruptured aneurysm releases blood into the spaces around the brain, called a subarachnoid hemorrhage (SAH) – a life-threatening type of stroke. Treatment options for aneurysms include observation, surgical clipping, coiling, and bypass.

[43] A circulatory anastomosis that supplies blood to the brain and surrounding structures. It is named after Thomas Willis.

[44] Arachnoiditis a pain disorder caused by the inflammation of the arachnoid, one of the membranes that surround and protect the nerves of the spinal cord.

aches, pain, weakness in the legs with a sensation of cold, and sometimes heat. In dreams P. often sees scenes that involve his legs in an unpleasant situation. For example, on the night of 13/IV in 1965 in a dream:

"I saw my stepmother, she was beaten on the head and then kicked barefoot on a snowy street. Ran long in the snow."

Upon awakening he had a headache and a cold feeling in the feet. The second patient, S., 26 years old, from the same institution (1965), suffering basal arachnoiditis, complained of a headache, feeling hot or cold in the face. In her dreams she very often saw the same scene:

"My temperature varies: then hot, then cold. I am standing in dirty water, for a very long time, and I continue to wash. I am in a hurry, because I am often late for work, and miss my train."

To reduce the dreams she went under hypnosis. The night immediately following this suggestion there was no dream, but a day or two later they reappeared.

Patients suffering from encephalitis: in five patients for 35 days, we recorded 29 dreams.

In the case of hypersomnia[45] dreams have been rare. The content of their dreams almost constantly are intense. Here are some examples of the patients and their dreams:

Patient P., 18 years (Institute Bekhterev, 1965), suffering from excessive sleepiness, said that he sleeps a lot, but does not remember a lot of dreams. In seven nights we recorded only one dream in which: *"lay on his bed in the chamber, I saw a lot of sleeping patients and I wanted to sleep."*

Another patient, 36 years old, from the same institution (1965), with the onset of severe headache, palpitations, chills, fear and

---

[45] The main symptom of hypersomnia is excessive daytime sleepiness (EDS), or prolonged nighttime sleep, which has occurred for at least 3 months prior to diagnosis

sharp weakness, had a disturbing night of sleep with an abundance of dreams. In one:

"I kill the dreaded snake, cut off his head, which was a small female head, I was afraid doing that"; in another dream: *"I saw the other patients in the same condition, their faces had turned blue and they had clenched jaws."*

Often in dreams from patients suffering from encephalitis the consequences of the disease were already visualized. We can say as a rule for these patients and patients without psychiatric violations for sleep theres is an abundance of unpleasant dreams, filled with symptoms associated with the disease.

For example, the patient S., 35 years old Institute Bekhterev 1965) eighteen years ago was affected by encephalitis, and suffers constant headaches and seizures. Sleeps light, dreams a lot, always unpleasant. For example, on the night of 1 / III 1965:

"my son was dead, lying on the cement floor, I cried, clutching hands behind his head, and then fell to the floor, it was cold, I was told: Arise, otherwise you catch a cold."

Upon awakening the patient experienced an alarming condition: headache, feeling cold all over, tears streaming from his eyes. Patients in the chamber next to the patient said that S. was crying and moaning during sleep.

Of certain interest is the the six-year observation of Vova and Glory V., twins who have craniotomy and a fusion of the skull bones, post subarachnoid spaces and vessels.

Despite their similarities, their behavior was often quite different. One angry, crying, another was calm and comforted the first one willing to participate with the doctors, the other would not, one regularly experiences vomiting, the other does not, one urinates in bed, the other not. They not only fell asleep and woke up sometimes at different time, but the dreams were different. For example, on the night of 4/XI 1966 Vova had a dream:

"were they were about to do surgery on his head, scared."

In the morning he complained of a headache. Glory this morning complained of a headache too, but did not remember dreams.

Glory on the night 9/XI 1966, in the dream *"written on the floor in the chamber."* On this night Glory really peed in the bed. Vova woke up dry and said he did not see dreams.

Very often we had dreams of patients with lesions of the spinal cord. We observed eight patients with tumors, arachnoiditis, lesions of different parts of the spinal cord and one with lumbosacral sciatica[46]. In 62 cases of sleep we recorded 60 dreams.

The majority of patients had cuts on the legs which complicated movement. The dream content was unpleasant to them, it frequently showed affected related areas on the spinal cord, leg and walking. Dreams scenes were often annoying: long walks in uncomfortable shoes or barefoot in the mud, expressing difficulty walking.

For example, the patient D. 58 years (Institute Bekhterev) suffers from arachnoiditis, spinal cord paresis of the legs, walks with a cane. The patient has many dreams were he often walks barefoot or in large boots in the mud. On the night of 22/I, 1965:

> *"walked barefoot through the mud, my feet were cold , I thought I might thinking catch a cold."* Upon awakening the patient feels pain in the feet and the lower legs are cold.

Only in a diseases with a long duration(10 - 17 years) dreams begin to express paresis and paralysis of the legs: often scenes appear associated with the inability to walk.

Thus, with the deterioration of the most diverse areas of the brain, although there were some changes in dreams, their extinction was observed. This fact contradicts the statement of Ullman, and M. Jouvet and other foreign authors of a special "center" for dreams.

---

[46] Sciatica (/saɪˈætɪkə/; sciatic neuritis, sciatic neuralgia, or lumbar radiculopathy is a set of symptoms including pain caused by general compression or irritation of one of five spinal nerve roots of each sciatic nerve—or by compression or irritation of the left or right or both sciatic nerves.

During the breakdown of the left hemisphere and particularly the left parietal area dreams are reported less frequently, than with a malfunctioning of the right area.[47]

With a significant destruction of the occipital lobe visions lose the brightness of coloring. Cortical hemianopia[48] appears direct in dreaming. When a patient completely loses sight during his lifetime due to damage to the eyes, optic nerves and tract frequency, dreams and their structure are almost unchanged whereas temporal cortical blindness is not marked in the dreams.

In the content of the dream there were some changes related to the localization process and clinical course of the disease, which gives grounds to use them for diagnostic purposes. Repeatedly having the same type of unpleasant dreams or nightmares, in which scenes express a head injury, ugly deformed heads, especially if the damage is associated with certain areas of the head, always should suggest the possibility of a tumor in the brain. Dreams can inform patients in topical diagnostics.

In conclusion of this chapter I want to focus on the issue of prodromal dreams, their onset, role and various events. In the literatu-

---

[47] Robert J. Hoss in *Science of Dreaming*: More recent evidence with more specific measurement tools, as noted in table 1, shows that it is more than just the right brain involved in dreaming, but rather various sections of the brain activating and de-activating that make the dream state more like right brain activity and less like left brain activity. This likely occurs because some of the more influential centers that are activated in the dream state, are specific to the right hemisphere, such as the right inferior parietal cortex. This is the visuospatial processing center of the brain perhaps involved in image and dream space construction. Also centers that are deactivated (such as the left parietal cortex, and dorsolateral prefrontal cortex) are responsible for processing functions that are typically associated with left hemisphere. Nofzinger found an increase in activation of the right hypothalamus and the right frontal cortex during REM sleep and a decrease in the left frontal cortex. Marquet found an increase in the right parietal cortex and decrease in the left during REM.

[48] Homonymous hemianopsia, also referred to as homonymous hemianopia is the loss of half of the field of view on the same side in both eyes. It occurs frequently in stroke, tumor and traumatic brain injuries, because of the manner in which the nasal nerve fibers from each eye cross as they pass to the back of the brain.

re there has been written a lot about those "prophetic dreams" (N. Groth, 1878, N. Popov, 1908, N. Rudnev, 1915, etc.), and in our current age some people still attach importance to such dreams (Damstra,

1953; Leavitt, 1957). It has been suggested that such prophetic dreams occur simultaneously in many people and are forerunners of great historical events. In particular, note that before the First World War (1914-1918), many people in Russia dreamed about the war with the Germans, before it actually happened.

Such dreams, according to our observations, also occurred a lot before the war of 1941-1945. Similar dreaming can be seen at the moment, and we understand their origin as explained below.

Questions that intrigue us, of course, are about the validity of such dreams because before the war there was a lot of news about various preparations that was noticed by broad sections of the people. Various types of propaganda (print, radio, etc.) have had impact on large numbers of people, so it is no wonder that many people under the influence of all these methods were inspired to dream about this terrible phenomenon. Consequently, such dreams are the result of certain influences from the outside, not a result of a force beyond the control of human life.

We saw that dreams can manifest what happened in the past or what is happening in the body during sleep under the influence of external or internal stimuli.

Dreams can be an investigating tool, to some extent, they may appear to measure only changes currently occurring to the body and the central nervous system. But they also reflect the past or with a breathtaking backdrop foreshadow events that can occur in the future. This applies also to disease. Dreams should be judged at the onset of a disease, it is possible to predict a disease which may occur in the future.

The following data is presented on how early dreams change in different diseases: Table 21:

Disease	# patients	How long until an explicit disease has changed the nature of dreams
Suppuration of furuncle	9	the night before
Acute appendicitis	3	the night before
Acute enterocolitis	7	the night before
Acute bronchitis and rhinitis airway	9	the night before
Acute gastritis and toxic infection	5	the night before
Flu	14	the night before
Angina	6	same night
Toothache	5	same night
Conjunctivitis hordeolyum	2	same night
Acute dysentery	3	In two cases before and one case in two days
Infectious psychosis	3	3-5 days
Delirium tremens	4	5-10 days
Botkin's disease	5	about a week
Typhoid	2	one week
Typhus	2	one week
Neuroses	18	from a week to several months
Chronic gastritis	2	about a month
Pulmonary tuberculosis	1	2 months

Hypertonic disease	5	2-3 months
Saint Martin's evil	5	a few monte
Intoxication psychosis	5	about a year
Brain tumors	15	from more than a year to a month
Total	130	

This are just the cases of diseases in which patients's clearly remembered a dream. A lot of patients miss the explicit manifestation or detection of diseases; because they failed to give an answer to the doctor questioning them about their dreaming, or they do not pay attention to their dreams. Sometimes a patient in the midst of his disease or during convalescence, no longer remembers how dreams have changed since the beginning of the disease.

However, the given data shows that quite often dreams change in a period of disease. It is interesting to note that the time preceding changes in dreams was often equal to the latent period of diseases. In infectious diseases like flu, sore throat, acute enterocolitis, dysentery, typhoid and typhus fever the dreams change right before the onset of the disease, this period seems somewhat to increase for diseases like neurosis and psychosis - and it seems even a longer period that dreams change before the onset of a brain tumor.

Of particular importance in the diagnosis of dreams are the slowly developing neuropsychiatric and internal diseases, such as brain tumors and tuberculosis of the lungs. Dreams may anticipate these diseases many months before their explicit manifestation.
Other authors cite a number of examples where a few days prior to the detection of the disease the content of dreams vary in a certain manner. Thus, even Galen described a case where one person dreamed that his leg was a stone, and in a few days this leg was paralyzed.

Lhermitte reported several such examples: in one case, a man had a dream that someone was bitten on the leg, a few days later on this leg anthrax[49] was found. In another case a man had a dream that he has a sore throat, and when he awakes this disease has arisen. A third example in the writings of Lhermitte is of a woman who often had frightening dreams: about fire and blood and always awakes in fear. She was diagnosed with rheumatic heart disease.

About nightmares with pronounced elements of fear M. Astvatsaturov writes: "For example, we recognize that if disturbing dreams with elements of fear of death are combined with sudden awakenings accompanied with unaccountable fear of death (angina pectoris sine dolore also known as Gairdner Disease), it may be the impact of a suspected heart disease in a period when there are no other subjective complaints indicating a disease".

There are many similar examples. Could it be possible by careful investigation to accurately judge the onset of the disease using dreams? Even if we could distinguish symptoms of diseases in dreams, they are not strictly consistent for all people; here we must distinguish between two groups of features: elements in dreams vary per person and dreams seem to indicate the last submissions relevant to the future[50]. In general, the most persistent symptoms associated with the feature of the disease may differ

---

[49] Anthrax is an acute disease caused by the bacterium *Bacillus anthracis*. Most forms of the disease are lethal, and it affects both humans and animals. There are now effective vaccines against anthrax, and some forms of the disease respond well to antibiotic treatment.

[50] MI Astvatsaturov (1935) proposed the idea of anatomical and physiological basis of dreams. He regarded dreams as manifestations of "somatic or mental viscero-switching", attaching great importance to this mechanism, the "Visual hillocks" of the brain.
The views expressed and the arguments were based on the idea of the relationship of dreams to the brain. Dreaming is a function of the brain, a kind of activity in the cerebral hemispheres time sleep. Today physiological science has a particular perspective to establish some link of dreams with various functional states of the cerebral cortex.

from person to person and depend on individual human traits that may occur in people differently.

At the beginning of this chapter, we explored the general features of dreams, connecting them with the peculiarities of various diseases. We observed that with one and the same disease in different people the content of dreams sometimes differ slightly.

It depends on the individual characteristics of the person: occupation, life, education, upbringing.

For example, a disease of the gastrointestinal tract is for most people, not familiar with medicine, related with meals, unpleasant tainted food, but medical professionals could still be alerted by the scenes of discontinuity that occur in these dreams because they express certain anatomical changes in the stomach and the affected intestine.

In angina for people without medical knowledge scenes appear expressing damage to the throat, and the doctor sees a sick throat, diphtheria and more morphological changes.

A warrior of ancient Greece with a skin disease of his legs dreamed that in this place

an arrow hit, while an officer during the Second World War with a similar disease. dreamed that a bullet hit him in the leg.

An officer that gets the flu, dreams about the war in the night before the explicit manifestations of his disease, and will get an injury to head and throat in his dream; a medical student falls ill and has a dream about an anatomical theater, ward, an electrician, gets the flu, and in his dream sees a bright electric light etc.

Dreams are dependent upon and will reflect education and other circumstances of people.

Some definitely argued that unpleasant dreams in which there are visual scenes of raw meat and/or blood, signify the beginning of common diseases, flu and other; and eating raw meat, raw fish, poor-quality food are associated with the gastrointestinal tract.

We observed N. who said that before the onset of a gastrointestinal disorder in her dreams there were often scenes related to food:

pickled mushrooms, and these dreams began to appear after she suffered a mushroom poisoning.

Others argue that dreaming about dirt, muddy water, hospital, pharmacy, doctors, as well as the dead, coffins, graves might be a symbol of a common disease that is starting. Often in such dreaming there are scenes of war, fight, fire, injury and surgery on different parts of the body, etc.

Common to these visual image and scenes is that they are somehow related to the disease or condition in the waking state. But different people link the disease with different visual images in the waking condition. As we have said, in N. a gastrointestinal disorder first appeared after eating Travleniya mushrooms, then viewing marinated mushrooms while awake causes discomfort; disgust, nausea, and so there is a conditioned reflex established between these dream scenes and the gastrointestinal disorder.

In our sample in a few people there was a definite relationship between raw meat in dreams and disease. In one person we recorded several dozen of such dreams. Questioning this person revealed that as a child he often heard stories relating raw meat to the disease from his father and mother. Being an impressionable boy, he was already alerted to this phenomenon, and therefore, already formed a conditioned reflex connecting disease with raw meat. In his youth he was sick for a long time, and in that time he often had dreams with an unpleasant character. They were rare at first, and then began to appear more and more. Scenes with raw, bloody meat as food. Eventually in the evening before going to bed, he was afraid to dream of raw meat, because those dreams always announced a deterioration of his health.

In this example, the first visual image of creates a tenuous connection between meat and the disease. In the waking state surrounding persons suggested the relationship so that what randomly appeared for the first time during the disease, was secured by the patient, and so firmly entrenched, that for the rest of his life disease was related to dreams about raw meat.

N., had suffered from a minor heart attack during the funeral of her husband, with subsequent cardiovascular disorders. N. often dreams about a cemetery, and often feels pain in her heart because she misses her husband. In this case the beginning of heart disease is associated with visual scenes of a cemetery and the deceased.

Mature woman S. told me that on the eve before she gets a disease (Influenza, catarrh of the upper respiratory tract), she usually dreams about bathing in muddy, dirty water, she sometimes even feels warm or cold water. And when she wakes up in such cases she usually is sweaty, and feels discomfort in her body: a weakness, feeling cold or heat.

Some people have dreams with very clear visual scenes, that directly express the disease.

For example, a young female doctor, who often suffers from eyelids diseases (hordeolyum), sees eyelids in a dream in this or the next day, she really has hordeolyum.

In nervousness the first hint of a beginning of disease are changing dreams with an emotional background, the appearance of anguish, anxiety, fear and nightmares, and

also unpleasant visual scenes associated with the life situation, facilitated the emergence of neurosis.

Long before obvious symptoms of mental illness are observed dreams show very troublesome scenes that evoke unpleasant emotions. Usually this synchronizes with events in waking life. At the beginning of a disease of the brain unpleasant emotions in dreams are combined with unpleasant visual scenes, often expressing head injury.

Thus, we see that in dreams really very often a disease manifests itself earlier than in the waking state. This is not due to the mysterious unconscious forces trying to prove the Freudian insights, but due to certain physiological characteristics of the brain during sleep and above all different depths of sleep inhibition in different functional zones, and a very high sensitivity visual parser.

Below, we elaborate on considering these features, now we only note that the sensitivity of the visual parser has a sensitivity thousands, millions, billions or more times higher then all other analyzers.

In the waking state at the beginning of a slowly evolving disease sometimes there will occur minor physical chemical and other changes. The irritation arising from this is signaled to the appropriate part of the brain. These signals may be below the thresholds of pain sensation. heat, pressure, and other peculiar feelings of a disease.

So people may not notice the onset of the disease, but during sleep the same irritation due

to a developing disease in a brain region may exceed the subconscious threshold. During the process of excitation, which is manifested by originally constructed visual scenes, connected with the disease. These are, in our opinion, early manifestations of the disease in the dream.

In this connection I would like to quote M.I. Astvatsaturov, who wrote: "*It may give a false impression that the dreams themselves, regardless of the content, may indicate the future. In fact, they only indicate the last submissions relevant to the future.*"

www.ingramcontent.com/pod-product-compliance
Lightning Source LLC
Chambersburg PA
CBHW031122180526
45160CB00005B/56/J